EZ

Math Workbook

Written for parents and secondary students by an experienced teacher / parent

- Covers math subjects for middle school & high school
- Complete and easy to follow examples
- Sample problems for each subject
- Learn to + - x and ÷ fractions & decimals; Percent and plane geometry

So "EZ" even your parents will understand it!

Louis Parchman

ISBN 978-1-63784-070-2 (paperback)
ISBN 978-1-63784-071-9 (digital)

Hawes & Jenkins Publishing
16427 N Scottsdale Road Suite 410
Scottsdale, AZ 85254
www.hawesjenkins.com

Printed in the United States of America

CONTENTS

CHAPTER 1

ROMAN NUMERALS

Today we see Roman Numerals on clocks, books, watches, movies, and even on our dollar bill. The Romans used a system of counting different from the system we use today. Below is a table of the values of their system. Notice that all 7 of the numbers begin with either 1 or 5. The Romans did not need 0, 2, 3, 4, 6 etc. By placing 2 X's together, they added their value and got 20.

Roman	Value
I	1
V	5
X	10
L	50
C	100
D	500
M	1000

The last 4 letters **L C D M** can be remembered by the sentence; "**L**ittle **C**ows **D**rink **M**ilk".

If several letters are together, figure their value by starting at the left and work toward the right. If a letter of smaller value is to the RIGHT of a letter of a larger value, add the values together; XVI = 10 + 5 + 1 = 16. CLX = 100 + 50 + 10 = 160. If a letter of smaller value is to the LEFT of a letter of larger value, subtract the smaller value from the larger value; IX = 10(X)—1(I) = 9. XL = 50(L)—10(X) = 40. CD = 500(D)—100(C). It is never necessary to have more than three of the same letter values together.

Always go to the next higher value; XXXX = 40, instead use XL (50 - 10 = 40), CCCC = 400, instead use CD (500 - 100 = 400).

Examples: IV = 5 - 1 = 4

VI = 5 = 1 = 6

XX = 10 + 10 = 20

XXX = 10 + 10 + 10 = 30

1

$$XIV = 10 + (5-1) = 14$$

$$XLV = (50 - 10) + 5 = 45$$

Note: VL IS NOT USED FOR 45, BECAUSE A SMALLER LETTER IN FRONT OF A LARGER LETTER MUST BE A NUMBER THAT BEGINS WITH 1.

Only 1, 10, 100 can be used in front of larger numbers.

ROMAN NUMERALS (CONTINUED)

XCV = (100 - 10) + 5 = 95 (not the letters VC)
MCMLXXV = 1000 = (1000 - 100) + 70 + 5 = 1975
LXXI = 50 + 20 + 1 = 71
XV = 10 + 5 = 15
XIX = 10 + (10 - 1) = 19
XLIX = (50 - 10) + (10 - 1) = 49

To change a decimal number to a Roman numeral begin with the first number on the left and work to the right.

$$\frac{173 = \text{CLXXIII}}{\begin{array}{l} 100 = \text{C} \\ 70 = \text{LXX} \\ \underline{3 = \text{III}} \\ 173 \end{array}}$$

For large numbers the Romans used a line drawn across the top of the numeral to mean that the value is multiplied 1000 times.

500	=	D
1,000	=	M
2,000	=	MM
5,000	=	$\overline{\text{V}}$
6,000	=	$\overline{\text{VI}}$
10,000	=	$\overline{\text{X}}$
50,000	=	$\overline{\text{L}}$
60,000	=	$\overline{\text{LX}}$
100,000	=	$\overline{\text{C}}$
1,000,000	=	$\overline{\text{M}}$

1.1 Problems: (Change to Decimal Numbers)

1.	XI	11.	XL	
2.	VIII	12.	XXIX	
3.	IV	13.	XCV	
4.	XIII	14.	XLII	
5.	XVII	15.	LII	
6.	XXXIII	16.	LX	
7.	XXVII	17.	LXXV	
8.	XXX	18.	LXXXVII	
9.	XXIV	19.	MCMI	
10.	XLIV	20.	MDCCLXXVI ($1 Bill)	

1.2 Change the following numbers to Roman Numerals;

1.	9	14.	99	
2.	11	15.	101	
3.	14	16.	400	
4.	16	17.	600	
5.	19	18.	The current year	
6.	20	19.	Your house number	
7.	21	20.	Your birth date	
8.	24	21.	900	
9.	26	22.	1,100	
10.	39	23.	535	
11.	49	24.	153,000	
12.	51	25.	62,545	
13.	76			

PLACE VALUE

In our system of counting numbers (the decimal system) every place has a different value. Every place value increases by multiples of 10's. Below is a table of place value to billions. The table could go on forever, but for our purposes, trillions will be sufficient.

								4	3	1	9
TRILLIONS	BILLIONS	HUNDRED MILLIONS	TEN MILLIONS	MILLIONS	HUNDRED THOUSANDS	TEN THOUSANDS	THOUSANDS	HUNDREDS	TENS	ONES	

The number 4,319 means: 4 thousands, 3 hundreds, 1 ten and 9 ones.

1.3 Problems: Write the word meanings of the following numbers;

1.   10,056—**1** ten thousand, **0** thousands, **0** hundreds, **5** tens & **6** ones.
2.   236
3.   791
4.   8059
5.   1101
6.   2007
7.   562
8.   490
9.   3123
10.   7199
11.   37
12.   99
13.   999
14.   29,099
15.   10,761
16.   123,679

NUMBERS—EVEN & ODD

There are 10 numerals in our system (0,1,2,3,4,5,6,7,8,9). Five are even (0,2,4,6,8) and five are odd (1,3,5,7,9).

Any number ending with an even numeral is called an <u>Even Number</u>. Any number ending with an odd numeral is called an <u>Odd Number</u>.

4561 ends in an odd numeral and therefore is an odd number. 324 ends in an even numeral and therefore is an even number.

1.4 Problems; Write odd or even by each of the following numbers:

1. 214
2. 111
3. 220
4. 9,996
5. 56,780
6. 21,013
7. 1,007
8. 2,999
9. 1,212
10. 765

COMMAS IN LARGE NUMBERS

Commas are placed in large numbers to help us read the numbers.

Place a comma (,) after every third numeral. Start at the FAR RIGHT (ones column) and work to the left placing a comma every 3 numbers.

20112 = 20,112
1234678 = 1,234,678

Problems;

1.5   Place commas in the correct place.

1.   56565   11. 98654
2.   797113   12. 1001
3.   8900   13. 77787771
4.   10000   14. 234671
5.   10000000   15. 88731109
6.   100000000   16. 43190678914
7.   711900718   17. 942286002
8.   29801236   18. 11171
9.   1239347   19. 7112461
10. 320761   20. 6667780000

DECIMAL POINT—LINING IT UP!

<u>All</u> numbers are understood to have a decimal point. If the decimal point is shown, finding it is simple. If the decimal point is not shown, it is always behind the last numeral.

$$4 = 4.$$
$$6{,}163 = 6{,}163.$$
$$147 = 147.$$

When adding or subtracting, always line up the decimal points in a straight line moving down.

$3 + 13 + 6{,}071 + 149$ rewritten should look like this:

$$
\begin{array}{r}
3. \\
13. \\
6{,}071. \\
\underline{147.}
\end{array}
$$

1.6 Problems: Rewrite the following problems vertically and then add or subtract.

1. $6 + 60 + 600$

2. $1 + 203 + 3{,}711 + 97$

3. $14 + 148 + 173$

4. $560 + 17 + 64$

5. $99 + 1$

6. $76 + 112 + 7$

7. $4{,}033 + 999 + 70$

8. $1 + 11 + 111 + 1{,}111$

9. 365 + 5630 + 2

10. 114 + 4,191 + 1 + 97

11. 367 - 24

12. 999 - 10

13. 1,000 - 999

14. 75 - 5

15. 83 - 38

16. 5,007 - 93

17. 4,100 - 909

18. 781 - 92

19. 5,321 - 971

20. 423 - 62

MATH SYMBOLS USING =, ≠, <, >

= equal
≠ not equal
< less than
> greater than

When comparing 2 sets of numbers it is often necessary to place certain math symbols between the two numbers. If the two sides are equal place a (=) between them. If they are not equal place a (≠) not equal between them.

$$3 = 2 + 1$$
$$6 - 4 = 1 + 1$$
$$4 \neq 6$$
$$5 + 3 \neq 9 + 1$$

1.7 Problems: Place an = or ≠ sign between the following sets of numbers:

1.   7____6 +1
2.   8 x 1____8
3.   9 + 0____Nine
4.   4 - 2____2 + 1
5.   16 + 1____17
6.   9 x 2____20 - 2
7.   1 + 10____11 x 0
8.   6 - 1____10 - 4
9.   15 + 15____50 - 30
10.   100 - 9____90 + 1
11.   31 + 2____42 x 1
12.   17 - 9____7

13.   100 + 1____100 x 1
14.   99 - 9____90
15.   44 x 1____50 - 6
16.   73 - 4____69 x 1
17.   7 x 1____17 x 0
18.   49 + 49____98 x 1
19.   23 + 32____60 - 5
20.   14 + 1____15 - 1

MATH SYMBOLS (CONTINUED)

When comparing inequalities (not equal) use < for "is less than"; and > for "is greater than". Always read the numbers from left to right as you would read a sentence. 4 > 3 is read "Four is greater than three". **Don't compare the number on the right to the number on the left!** Hint: The POINT of the arrow will always point toward the SMALL number. 9 < 10, 11 > 10, 5 > 4.

1.8 Problems: Place < or > between the values:

1. 11 _____ 12
2. 12 _____ 11
3. 6 + 1 _____ 8
4. 9 - 1 _____ 7
5. 2 x 3 _____ 5
6. 14 _____ 13 + 3
7. 100 _____ 90 x 2
8. 5 + 5 + 4 _____ 10
9. 19 _____ 20 - 2
10. 16 x 1 _____ 17

1.1 Addition of whole numbers—(Review)

1. $\begin{array}{r} 67 \\ + 11 \\ \hline \end{array}$ 5. $\begin{array}{r} 7124 \\ + 7124 \\ \hline \end{array}$

2. $\begin{array}{r} 94 \\ + 36 \\ \hline \end{array}$ 6. $\begin{array}{r} 932 \\ + 979 \\ \hline \end{array}$

3. $\begin{array}{r} 792 \\ +123 \\ \hline \end{array}$ 7. $\begin{array}{r} 3693 \\ + 4697 \\ \hline \end{array}$

4. $\begin{array}{r} 496 \\ + 934 \\ \hline \end{array}$ 8. $\begin{array}{r} 73650 \\ + 97929 \\ \hline \end{array}$

9. 37
    73
    11
    99
    +12
    ─────

11. 1000
     1001
    + 999
    ─────

10. 143
     632
    + 117
    ─────

12. 1000 + 50 + 1000 + 50 =

SUBTRACTION OF WHOLE NUMBERS

Subtraction, often referred to as "take away", is easy. Subtraction is the opposite of addition. 5 - 2 = 3 means the 2 is subtracted from 5 leaving 3. Think: what number must be added to 2 to equal 5? Answer; 3.

When checking subtraction, add the subtrahend (bottom number) to the answer and you will get the minuend (top number).

$$
\begin{array}{r} 8 \\ -\ 3 \\ \hline 5 \end{array}
\quad \text{check} \quad
\begin{array}{r} 5 \\ +\ 3 \\ \hline 8 \end{array}
$$

It is recommended that you always check your subtraction

When subtracting, it is sometimes necessary to "borrow" from the next digit.

$$
\begin{array}{r} {}^{0}\ {}^{9} \\ \cancel{1}\ \cancel{0}\ 1 \\ -\ \ 98 \\ \hline \end{array}
$$

You cannot take 8 from 1 so "borrow" one from the 10 making it 9 and making the 1 an 11.

1.10 Problems:

1. 19	2. 21	3. 49	4. 74
$-\ 11$	$-\ 20$	$-\ 13$	$-\ 51$

5. 99	6. 75	7. 29	8. 69
$-\ \ 1$	$-\ 35$	$-\ 19$	$-\ 33$

9. 14	10. 123	11. 345	12. 627
$-\ 13$	$-\ 12$	$-\ 214$	$-\ 163$

13. 718	14. 479	15. 583	16. 698
- 207	- 340	- 123	- 451

17. 964	18. 1903	19. 2711	20. 4319
- 504	- 903	- 2011	- 1000

21. 432	22. 691	23. 731	24. 730
- 320	- 580	- 411	- 710

25. 999
   - 110

1.11 Subtracting with borrowing. Subtract and check:

```
              0
1.    ︢7    check    9          4.      40
     - 9           + 8                 - 38
      8            17


                                 5.      33
                                        - 28
2.    63
     -18


                                 6.      70
3.    72                                - 61
     - 9
```

7.   309
      - 99

14. 1768
     - 769

8.   431
   - 288

15. 1000
     - 999

9.   1000
   - 798

16. 32002
    - 6099

10.   143
    - 94

17. 362
   - 199

11. 1654
   - 664

18. 1200
   - 300

12. 744
   - 348

19. 3624
   - 3525

13. 1297
   - 209

20. 1009
   - 899

MULTIPLICATION OF WHOLE NUMBERS

To learn multiplication, first memorize the table below:

	1	2	3	4	5	6	7	8	9	10
1	1	2	3	4	5	6	7	8	9	10
2	2	4	6	8	10	12	14	16	18	20
3	3	6	9	12	15	18	21	24	27	30
4	4	8	12	16	20	24	28	32	36	40
5	5	10	15	20	25	30	35	40	45	50
6	6	12	18	24	30	36	42	48	54	60
7	7	14	21	28	35	42	49	56	63	70
8	8	16	24	32	40	48	56	64	72	80
9	9	18	27	36	45	54	63	72	81	90
10	10	20	30	40	50	60	70	80	90	100

6 x 8: Go across the top to the number 6, then go down to the number 8. Where they intersect is at the number 48. 6 x 8 = 48.

When multiplying larger numbers multiply the multiplier (bottom number) times the multiplicand (top number).

			1
Step 1.	2 x 5 = 10;	put a zero and carry the one.	45 (multiplicand)
Step 2.	2 x 4 = 8;	plus the one you carried, equal 9.	x 2 (multiplier)
			90

To multiply 2-digit numbers follow the steps below:

```
      29
   x  46
    174
    116
   1334  (product)
```

1. 6 x 9 = 54; put down the 4 and carry the 5
2. 6 x 2 = 12, plus the 5 you carried = 17
3. Indent to the left one digit space.
4. 4 x 9 = 36; put down the 6 and carry the 4
5. 4 x 2 = 8 + 3 (the 3 was carried over) = 11.
6. Add the partial products.

MULTIPLICATION OF WHOLE NUMBERS (CONTINUED)

Remember to INDENT one place to the left in the second row and every row after that.

$$
\begin{array}{r}
319 \\
\times\ 128 \\
\hline
2552 \\
638 \\
913 \\
\hline
40,832
\end{array}
$$

1.12 Problems; Multiply the following problems:

1.   10 x 9 =

2.   11 x 6 =

3.   12 x 10 =

4.   6 x 9 =

5.   14 x 9 =

6.   17 x 3 =

7.   12
      x 4

8.   12
      x 13

9.   132
      x 14

10.   324
      x 14

11.   44
      x 21

12.   84
      x 5

13.   62
      x 15

MULTIPLICATION OF WHOLE NUMBERS (CONTINUED)

14.   12
    x 7

15.   129
    x 6

16.   14
    x 14

17.   503
    x 9

18.   437
    x 14

19.   24
    x 10

20.   102
    x 40

21.   612
    x 19

22.   473
    x 97

23.   1234
    x 121

24.   675
    x 14

25.   333
    x 33

26.   111
    x 111

27.   125
    x 10

28.   69
    x 34

29.   505
    x 12

30.   374
    x 120

DIVISION OF WHOLE NUMBERS

When dividing, you are trying to find out how many times the divisor will go into the dividend. The answer is called the quotient.

$$\text{divisor}—5/\overline{10}\underset{\text{—dividend}}{\overset{2\text{—quotient}}{}}$$

It is sometimes necessary to carry in division.

Step 1. Divide 5 into 6 = 1.

2. Multiply 1 x 5 = 5.

3. Subtract 5 from 6 = 1.

4. Since 5 will not go into 1, bring down the next number (5)

5. Divide 5 into 15 = 3.

6. Multiply 3 x 5 = 15.

```
        13
   5/  65
      - 5
       15
       15
        0
```

```
        12 R 3
   5/  63
      -  5
        13
        10
         3 - remainder
```

1.13 Problems; Divide the following:

1. $2/\overline{6}$

2. $3/\overline{24}$

3. $4/\overline{40}$

4. $5/\overline{65}$

5. $6/\overline{18}$

6. $7/\overline{49}$

7. $8\overline{)48}$

13. $6\overline{)546}$

19. $7\overline{)100}$

8. $9\overline{)72}$

14. $4\overline{)92}$

20. $8\overline{)42}$

9. $10\overline{)90}$

15. $5\overline{)132}$

10. $11\overline{)88}$

16. $6\overline{)123}$

11. $4\overline{)248}$

17. $2\overline{)13}$

12. $9\overline{)189}$

18. $6\overline{)61}$

SHORTCUTS TO MULTIPLYING AND DIVIDING BY 10, 100 AND 1000:

There is a shortcut to multiplying by multiples of 10. This shortcut will work for any number, as long as it is multiplied by a multiple of 10 (10, 100, 1000 etc.) Move the decimal point in the number to the **RIGHT** the number of places as there are zeros in the number you are multiplying. (Add zeros if necessary)

17.34 x 10 = 173.4
.1734 x 100 = 17.34
75 x 1000 = 75000.

The shortcut in division works just the opposite. Move the decimal point to the **LEFT**, instead of the right. In division, we are making numbers smaller, in multiplication we are making numbers larger.

21.7 ÷ 10 = 2.17
323 ÷ 10 = 32.3
.123 ÷ 100 = .00123
8.3 ÷ 10 = .83

1.14 Problems; multiply or divide using the **shortcut** method:

1. .7 ÷ 10	10. 8.7 ÷ 100	19. 7.1 ÷ 10
2. .62 x 10	11. 77 ÷ 10	20. 3.7 ÷ 100
3. 11.2 ÷ 100	12. 7.7 ÷ 10	21. 1121 ÷ 1000
4. 17 ÷ 10	13. 63 ÷ 100	22. 22 ÷ 10
5. 4 x 100	14. .19 x 10	23. 22 ÷ 100
6. .42 x 10	15. 4.3 ÷ 1000	24. 2.2 ÷ 1000
7. 1.11 x 10	16. 1.7 ÷ 100	25. 6 ÷ 10
8. .731 x 100	17. .425 x 100	26. 7.3 x 10
9. 6.11 x 1000	18. .41 x 100	27. .121 x 100

RULES FOR DIVISIBILITY

To determine if a number can be divided by 2 thru 10, there are different rules to apply. Rules:

A number can be divided by 2 if the last digit is an even number (0,2,4,6,8). Examples: 30, 72, 54 or 178

A number can be divided by 3 if the sum of its digits can be divided by 3.

Examples:	123 (1 + 2 + 3 = 6)	6 is divisible by 3
	201 (2 + 0 + 1 = 3)	3 is divisible by 3
	936 (9 + 3 + 6 =18)	18 is divisible by 3

A number can be divided by 4 if the **last 2 digits** can be divided by 4.

Examples:	10**84** (84 is divisible by 4)
	3**48** (48 is divisible by 4)

A number can be divided by 5 if it ends in 5 or 0.

Examples: 10**5**, 63**0** or 73**5**

A number can be divided by 6 if the number can be divided by both 2 **AND** 3.

Examples:	132 (divisible 2-even-and 3, sum of the digits = 6).
	318 (divisible 8-even-and 3, sum of the digits =12).
	474 (divisible 4-even-and 3, sum of the digits =15).

The rule for the number 7 is very complicated, so it is easier to divide the long way. A number can be divided by 8 if the **last** 3 digits are divisible by 8.

Examples: 2**136**, 2**216**, 4**312**

A number can be divided by 9 if the sum of its digits are divisible by 9.

Examples:	1053 (1 + 0 + 5 + 3 = 9)
	7965 (7 + 9 + 6 + 5 = 27)
	6813 (6 + 8 + 1 + 3 = 18)

A number can be divided by 10 if the last digit is a zero. Examples: 15<u>0</u>, 61<u>0</u> and 47<u>0.</u>

1.15 Problems: Write which numbers (<u>2 thru 10</u>) can be divided into the following numbers:

1. 16	8. 7875	15. 2041
2. 15	9. 564	16. 64
3. 25	10. 2084	17. 70
4. 40	11. 448	18. 36
5. 100	12. 3123	19. 44
6. 21	13. 1311	20. 180
7. 3136	14. 741	

USING ZERO IN THE FOUR OPERATIONS

The number 0 has no value but does hold place. 109 means: 1 hundred, 0 tens and 9 ones.

If you add or subtract zero (by itself) nothing happens. 3 + 0 = 3, 3 - 0 = 3. When you multiply any number by zero the answer is always 0. 9 x 0 = 0, 7 x 0 = 0, 19 x 0 = 0 and 123 x 0 = 0.

When dividing by zero the answer has no meaning, or the answer is UNDEFINED in mathematics.

We know 15 ÷ 5 = 3 because 5 x 3 = 15, 15 ÷ 0 = ? Using this example we can see that no number placed in the answer and multiplied by 0 would equal 15. However; when zero is divided BY a number, the answer is 0, not undefined. 0 ÷ 5 = 0 because 0 x 5 = 0.

1.16 Problems: work the following;

1. 16 + 0 =	6. 0 + 4 =	11. 0 ÷ 14 =
2. 17 - 0 =	7. 125 - 0 =	12. 14 ÷ 0 =
3. 17 + 0 =	8. 13 + 0 =	13. 19 + 0 =
4. 4 ÷ 0 =	9. 70 + 0 =	14. 0 + 19 =
5. 0 ÷ 4 =	10. 175 - 0 =	15. 21 + 0 =

EXPONENTS

To write large numbers it is easier to write them in the form of exponents. In the number 10^3, 10 is called the base and 3 is the power to which 10 is raised and is called the exponent. The number 3 tells us to multiply 10 three times. 10 x 10 x 10 = (10 x 10 =100 x 10 = 1000).

10^3 is read 10 "cubed"
5^2 is read 5 "squared" 7^4
is read 7 "to the 4th"
2^3 = 2 x 2 x 2 = 8
5^1 = 5 x 1 = 5
6^0 = 1*

*Note any number raised to the zero power equals 1.
Any number raised to the first power equals the number itself.

1.17 Fill in the spaces in the table below:

Numbers	Power	Base	Value
2^3	3	2	8
6^1			
7^0			
16^2			
3^2			
4^2			
5^4			
6^2			
8^1			
10^1			
12^2			
11^2			
173^0			
	3	6	
	5	3	
	9	1	
	1	15	
	0	192	
	7	2	
3^3			

TEST CHAPTER 1—WHOLE NUMBERS

1. 6 + 31 + 173 =

3. 75
    19
    43
    + 11

4. 112
    129
    733
    + 454

5. 226
    739
    + 432

2. 19 + 21 + 3 + 400 =

Subtract:

6. 17 - 9 =

8. 93
    - 81

9. 705
    - 316

10. 836
    - 369

7. 75
    - 25

Multiply:

11. 11 x 9 =

13. 63
    x 94

14. 101
    x 37

15. 44
    x 90

12. 17 x 13 =

Divide:

16. 16 ÷ 2 =

18. $15\overline{)165}$

19. $12\overline{)146}$

17. $21\overline{)105}$

20. $43\overline{)516}$

TEST CHAPTER 1—WHOLE NUMBERS (CONTINUED)

<u>Change to regular numbers:</u>

21. XIX   22. XLV

<u>Change to Roman Numerals:</u>

23. 90   24. 40

<u>Write in Numbers:</u>

25. Three Thousand, Two Hundred, Seven

Chapter 1.1

1	11	11	40
2	8	12	29
3	4	13	95
4	13	14	42
5	17	15	52
6	23	16	60
7	27	17	75
8	30	18	87
9	24	19	1901
10	44	20	1776

Chapter 1.2

1	IX	14	XCIX
2	XI	15	LI
3	XIV	16	CD
4	XVI	17	DC
5	XIX	18	MMXV (2015)
6	XXI	19	YOUR HOUSE #
7	XXI	20	YOUR DoB
8	XXIV	21	CM
9	XVI	22	MC
10	XXXIX	23	DXXXV
11	XLI	24	CLMMM
12	LI	25	LXMMDXLV
13	XLLVI		

Chapter 1.3

1	Example
2	2 Hundreds, 3 Tens, 6 Ones
3	7 Hundreds, 9 Tens, 1 Ones
4	8 Thousands, 0 Hundreds, 5 Tens, 9 Ones
5	1 Thousand, 1 Hundred, 0 Tens, 1 Ones
6	2 Thousands, 0 Hundreds,) Tens, 7 Ones
7	5 Hundreds, 6 Tens, 2 Ones
8	4 Hundreds, 9 Tens, 0 Ones
9	3 Thousands, 1 Hundred, 2 Tens, 3 Ones
10	7 Thousands, 1 Hundred, 9 Tens, 9 Ones
11	3 Tens, 7 Ones

12	9 Tens, 9 Ones
13	9 Hundreds, 9 Tens, 9 Ones
14	2 Ten Thousands, 9 Thousands, 0 Hundreds, 9 Tens, 9 Ones
15	1 Ten Thousands, 0 Thousands, 7 Hundredths, 6 Tens, 1
16	1 Hundred Thousand, 2 Ten Thousands, 3 Thousands, 6 Hundreds, 7 Tens, 9 Ones

Chapter 1.4

1	Even	6	Odd
2	Odd	7	Odd
3	Even	8	Odd
4	Even	9	Even
5	Even	10	Odd

Chapter 1.5

1	56,565	11	98,654
2	797,113	12	1,001
3	8,900	13	77,787,771
4	10,000	14	234,671
5	10,000,000	15	88,731,109
6	100,000,000	16	43,190,678,914
7	711,900,718	17	942,286,002
8	29,801,236	18	11,171
9	1,239,347	19	7,112,461
10	320,761	20	6,667,780,000

Chapter 1.6

1)
```
        6
       60
   +  600
      666
```

6)
```
       76
      112
   +    7
      195
```

2)
```
        1
      203
     3711
   +   97
     4012
```

7)
```
     4033
      999
   +   70
     5102
```

3)
```
       14
      148
   +   73
      235
```

8)
```
         1
        11
       111
   +  1111
      1234
```

4)
```
      560
       17
   +   64
      641
```

9)
```
      365
     5630
   +    2
     5997
```

5)
```
       99
   +    1
      100
      641
```

10)
```
      114
     4191
        1
   +   97
     4403
```

11)	367		16)	5007
	- 24			- 93
	343			4914

12)	999		17)	4100
	- 10			- 909
	989			3191

13)	1000		18)	781
	- 999			- 92
	1			689

14)	75		19)	5321
	- 5			- 971
	70			4350

15)	83		20)	423
	- 38			- 62
	45			361

Chapter 1.7

1   $7 = 6 + 1$
2   $8 \times 1 = 8$
3   $9 + 0 = 9$
4   $4 - 2 < 2 + 1$
5   $16 + 1 = 17$
6   $9 \times 2 = 20 - 2$
7   $1 + 10 > 11 \times 0$
8   $6 - 1 < 10 - 4$
9   $15 + 15 > 50 -$
10 30
11 $100 - 9 = 90 + 1$

12   $31 + 2 < 42 \times 1$
13   $17 - 9 > 7$
14   $100 + 1 > 100 \times$
15   1
16   $99 - 9 = 90$
17   $44 \times 1 = 50 - 6$
18   $73 - 4 = 69 \times 1$
19   $7 \times 1 > 17 \times 0$
20   $49 + 49 = 98 \times 1$
21   $23 + 32 = 60 - 5$
22   $14 + 1 > 15 - 1$

Chapter 1.8

1	11 < 12
2	12 > 11
3	6 + 1 < 8
4	9 - 1 > 7
5	2 x 3 > 5
6	14 < 13 + 3
7	100 < 90 x 2
8	5 + 5 + 4 > 10
9	19 > 20 - 2
10	16 x 1 < 17

Chapter 1.9

1	78
2	130
3	915
4	1,430
5	14,248
6	1,911
7	99
8	171,579
9	4,784
10	763
11	8,390
12	2,100

Chapter 1.10

1	8
2	1
3	32
4	23
5	98
6	40
7	10
8	36
9	1
10	111
11	131
12	464
13	511
14	139
15	460
16	247
17	460
18	1,000
19	700
20	3,319
21	112
22	111
23	320
24	20
25	889

Chapter 1.11

1	8
2	45
3	63
4	2
5	5
6	9
7	210
8	143
9	202
10	49
11	990
12	396
13	1088
14	999
15	1
16	25,903
17	163
18	900
19	99
20	110

Chapter 1.12

1	90	16	196
2	66	17	4,527
3	120	18	6,118
4	454	19	240
5	126	20	4,080
6	51	21	11,628
7	48	22	45,881
8	156	23	149,314
9	1,848	24	9,450
10	4,536	25	10,989
11	924	26	12,321
12	420	27	1,250
13	930	28	2,346
14	84	29	6,060
15	774	30	44,880

Chapter 1.13

1	3	11	62
2	8	12	21
3	10	13	91
4	13	14	23
5	3	15	26 R2
6	7	16	205
7	6	17	6 R1
8	8	18	10 R1
9	9	19	14 R2
10	8	20	5 R2

Chapter 1.14

1	0.07	15	0.0043
2	6.2	16	0.017
3	0.112	17	42.5
4	1.7	18	41
5	400	19	0.71
6	4.2	20	0.037
7	11.1	21	1.121
8	73.1	22	2.2
9	6,110	23	0.22
10	0.087	24	0.0022
11	7.7	25	0.6
12	0.77	26	73
13	6.3	27	12.1
14	1.9		

Chapter 1.15

1	2,4,8	11	2,4,8
2	3,5	12	3,9
3	5	13	3
4	2,4,8,10	14	3
5	2,4,5,10	15	-
6	3,7	16	2,4,8
7	2,4,8	17	2
8	3,5	18	2,3,4,6,9
9	2,3,4,6	19	2
10	2,4	20	2,3,4,5,6,9

Chapter 1.16

1 16
2 17
3 17
4 0
5 0
6 4
7 125
8 13
9 13
10 175
11 0
12 U
13 19
14 19
15 21

Chapter 1.17

Numbers	Power	Base	Value
2^3	3	2	8
6^1	1	6	6
7^0	0	7	1
16^2	2	16	256
3^2	2	3	9
4^2	2	4	16
5^4	4	5	625
6^2	2	6	36
8^1	1	8	8
10^1	1	10	10
12^2	2	12	144
11^2	2	11	121
173^0	0	173	1
6^3	3	6	216
3^5	5	3	405
1^9	9	1	1
15^1	1	15	15
192^0	0	192	1
2^7	7	2	128
3^3	3	3	27

Chapter 1
TEST

1	210	14	3,737
2	443	15	3,960
3	148	16	8
4	1,428	17	5
5	1,397	18	11
6	8	19	12R2
7	50	20	12
8	12	21	9
9	389	22	45
10	467	23	XC
11	99	24	XL
12	51	25	3,207
13	5,922		

CHAPTER 2

PRIME AND COMPOSITE NUMBERS

A number that has **exactly** 2 factors (one and the number itself) is called a prime number. A "factor" is a number that will divide into the number being divided without a remainder. The number 12 has six factors: 1, 2, 3, 4, 6 and 12.

A number that has more than 2 factors is called a composite number. EVERY number has at least 2 factors. If it has any more than 2, then it is a composite number.

Some examples of prime numbers are: 2, 3, 5, 7 and 11.

Some examples of composite numbers are: 4, 8, 20, 100 and 15.

Two is the lowest prime number and is also the only even number that is prime. All other prime numbers are odd.

Caution: This does not mean all odd numbers are prime! For example, 15 is odd but is not a prime number because it has more than 2 factors (1, 3, 5 and 15).

2.1 Problems:

1.   How many prime numbers are even?

2.   How many prime numbers are there between 1 and 100?

SIEVE OF ERATOSTHENES

About 2000 years ago a Greek mathematician named Eratosthenes used a method to sieve out (mark out) all the whole numbers that were NOT prime. See if you can do this for the numbers 1 to 100. 1 is not a prime number so it is marked out. The first row is done for you.

1	2	3	4	5	6	7	8	9	10
11	12	13	14	15	16	17	18	19	20
21	22	23	24	25	26	27	28	29	30
31	32	33	34	35	36	37	38	39	40
41	42	43	44	45	46	47	48	49	50
51	52	53	54	55	56	57	58	59	60
61	62	63	64	65	66	67	68	69	70
71	72	73	74	75	76	77	78	79	80
81	82	83	84	85	86	87	88	89	90
91	92	93	94	95	96	97	98	99	100

THE MEANING OF A FRACTION

A fraction has two parts. The top number tells us the number of parts and is called the <u>numerator</u>. The bottom number is called the <u>denominator</u>. This can be remembered by recalling the denominator is the number "down" and both denominator and down start with the same letter, <u>D</u>. Understanding fractions and their four operations (+, -, x, ÷) is not hard if we learn the rules and get a good understanding of what they mean.

Proper and Improper Fractions

A **proper** fraction is one that has a value of <u>less</u> than one. In other words, the numerator (top number) will <u>always</u> be smaller than the denominator (bottom number). Here are some examples of proper fractions:

$$\frac{1}{2} \qquad \frac{3}{7} \qquad \frac{19}{20} \qquad \frac{18}{25}$$

An improper fraction is <u>equal</u> to **or** <u>more</u> than one. The numerator will be equal to or more than the denominator. Some examples are:

$$\frac{7}{7} \qquad \frac{9}{3} \qquad \frac{10}{1} \qquad \frac{1}{1} \qquad \frac{6}{6} \qquad \frac{11}{4} \qquad \frac{3}{2}$$

Any fraction whose numerator and denominator are the same number is equal to 1.

$$\frac{7}{7} \qquad \frac{1}{1} \qquad \frac{16}{16} \qquad \frac{5}{5}$$

Comparing Fractions

Often it is necessary to compare 2 different fractions to determine if they are equal. A method called <u>cross multiplying</u> is used and is very easy. Are the fractions 2/3 and 6/9 equal? Yes.

Think: $\frac{2}{3} = \frac{6}{9}$ multiply the top of the fraction times the bottom of the other fraction and then do the same

to the other side. (the above example)

$\frac{2}{3} \diagup\!\!\!\!\diagdown \frac{6}{9}$ 2 x 9 = 18   Is <u>3</u> equal to <u>5</u>? No. $\frac{3}{4} \diagup\!\!\!\!\diagdown \frac{5}{10}$ 3 x 10 = 30

               3 x 6 = 18   4 10 4 x 5 = 20

2.3 COMPARING FRACTIONS (continued)

1. $\dfrac{1}{2}$ $\qquad$ $\dfrac{3}{6}$

2. $\dfrac{2}{5}$ $\qquad$ $\dfrac{6}{15}$

3. $\dfrac{4}{5}$ $\qquad$ $\dfrac{1}{2}$

4. $\dfrac{3}{7}$ $\qquad$ $\dfrac{10}{12}$

5. $\dfrac{6}{9}$ $\qquad$ $\dfrac{1}{9}$

6. $\dfrac{2}{3}$ $\qquad$ $\dfrac{3}{4}$

7. $\dfrac{1}{4}$ $\qquad$ $\dfrac{25}{100}$

8. $\dfrac{5}{6}$ $\qquad$ $\dfrac{10}{12}$

9. $\dfrac{5}{8}$ $\qquad$ $\dfrac{10}{16}$

10. $\dfrac{2}{3}$ $\qquad$ $\dfrac{8}{12}$

11. $\dfrac{1}{3}$ $\qquad$ $\dfrac{3}{9}$

12. $\dfrac{1}{4}$ $\qquad$ $\dfrac{2}{12}$

13. $\dfrac{1}{4}$ $\qquad$ $\dfrac{5}{16}$

14. $\dfrac{1}{2}$ $\qquad$ $\dfrac{3}{8}$

15. $\dfrac{1}{9}$ $\qquad$ $\dfrac{10}{90}$

16. $\dfrac{1}{3}$ $\qquad$ $\dfrac{1}{4}$

17. $\dfrac{3}{4}$ $\qquad$ $\dfrac{9}{12}$

18. $\dfrac{4}{5}$ $\qquad$ $\dfrac{5}{6}$

19. $\dfrac{1}{10}$ $\qquad$ $\dfrac{11}{1}$

20. $\dfrac{7}{9}$ $\qquad$ $\dfrac{9}{7}$

REDUCING FRACTIONS

Always reduce the answer in fractions, if possible. To see if a fraction can be reduced, try to think of a number that can be divided <u>into</u> the top **and** bottom number of the fraction.

4/12 can be reduced to 1/3 because 4 will divide into both 4 and 12. Remember the number you try must go into both the top **and** the bottom number of the fraction. 5/12 cannot be reduced even though 2, 3, 4, 6 & 12 will go into 12; none of these numbers will go into 5.

Examples: $\underline{6}$ reduce by dividing 3 into 6 and 9. $\dfrac{2}{3}$
9

$\underline{10}$ reduce by dividing 5 into 10 and 15. $\dfrac{2}{3}$
15

$\underline{2}$ reduce by dividing 2 into 2 and 6. $\dfrac{1}{3}$
6

If you can think of more than one number that will divide into a fraction, use the highest number. This will eliminate steps in reducing. 12/24 could be reduced by dividing 2, 3, 4, 6 or 12 into 12 and 24. You should use 12 and 12/24 reduces to 1/2.

2.4 Problems; Reduce the following fractions to lowest terms:

1. $\dfrac{6}{9}$

2. $\dfrac{5}{10}$

3. $\dfrac{8}{12}$

4. $\dfrac{10}{30}$

5. $\dfrac{4}{8}$

6. $\dfrac{25}{30}$

7. $\dfrac{10}{50}$

8. $\dfrac{2}{10}$

9. $\dfrac{2}{4}$

10. $\dfrac{4}{6}$

11. $\dfrac{6}{8}$

12. $\dfrac{6}{10}$

13. $\dfrac{3}{12}$

14. $\dfrac{4}{12}$

15. $\dfrac{9}{12}$

16. $\dfrac{5}{15}$

17. $\dfrac{10}{15}$

18. $\dfrac{3}{15}$

19. $\dfrac{2}{20}$

20. $\dfrac{14}{20}$

CHANGING AN IMPROPER FRACTION TO A WHOLE OR MIXED NUMBER

To change an improper fraction to a mixed number, divide the denominator (bottom number) into the numerator (top number). 15/3 = 5 (or 3 divided into 15 = 5). 18/2 = 9 (2 into 18 = 9).

If you have a remainder, express it as a fraction. The remainder is the numerator, the denominator stays the same.

$$22/7 = 7 \overline{)\,22\,}^{\,3} = 3\tfrac{1}{7}$$
$$\underline{21}$$
$$1 \text{ R.}$$

2.5 Problems; Change the improper fractions below to whole or mixed numbers:

1. $\dfrac{18}{3}$

2. $\dfrac{20}{4}$

3. $\dfrac{32}{5}$

4. $\dfrac{21}{6}$

5. $\dfrac{18}{8}$

6. $\dfrac{10}{4}$

7. $\dfrac{33}{11}$

8. $\dfrac{10}{4}$

9. $\dfrac{15}{10}$

10. $\dfrac{7}{2}$

CHANGING A MIXED NUMBER TO A FRACTION

To change a mixed number to a fraction, multiply the bottom part of the fraction times the whole number. Then **add** the numerator (top number) to your answer and this gives you the new top number. The bottom number stays the same.

$$6\frac{3}{4} = \frac{4 \times 6 = 24 + 3 = 27}{4} = \frac{27}{4}$$

$$4\frac{2}{3} = \frac{3 \times 4 = 12 + 2 = 14}{3} = \frac{14}{3}$$

2.6   Problems; Change the mixed numbers below to improper fractions:

1. $1\frac{1}{2}$

2. $2\frac{1}{2}$

3. $4\frac{3}{4}$

4. $10\frac{1}{9}$

5. $3\frac{1}{2}$

6. $5\frac{2}{3}$

7. $7\frac{3}{7}$

8. $9\frac{3}{2}$

9. $11\frac{1}{3}$

10. $2\frac{2}{6}$

11. $3\frac{3}{7}$

12. $5\frac{2}{5}$

13. $4\frac{2}{3}$

14. $3\frac{1}{2}$

15. $2\frac{1}{2}$

16. $1\frac{2}{3}$

17. $7\frac{4}{5}$

18. $6\frac{7}{10}$

19. $1\frac{5}{6}$

20. $2\frac{3}{4}$

FINDING THE MISSING NUMBER IN FRACTIONS

Using the cross multiplication process, we can determine the missing number when two fractions are compared. For example:

$\frac{?}{3} \times \frac{4}{6}$, we know that 3 x 4 = 12, so just think of a number times 6 that equals

12—answer <u>2</u>. $\qquad\qquad$ $\frac{2}{3} = \frac{4}{6}$ $\qquad\qquad\qquad$ 3 x 4 =12

$\qquad\qquad\qquad\qquad\qquad\qquad\qquad\qquad\qquad\qquad$ 2 x 6 = 12.

A missing number can be in any of the four parts of the two fractions.

$\qquad$ $\frac{?}{4} = \frac{5}{10}$ $\qquad\qquad$ 4 x 5 = 20 $\qquad\qquad\qquad$ Answer <u>2</u>
$\qquad\qquad\qquad\qquad$ what times 10 = 20;

$\qquad$ $\frac{9}{?} = \frac{18}{2}$ $\qquad\qquad$ 9 x 2 = 18 $\qquad\qquad\qquad$ Answer <u>1</u>
$\qquad\qquad\qquad\qquad$ what times 18 = 18;

$\qquad$ $\frac{6}{2} = \frac{?}{3}$ $\qquad\qquad$ 6 x 3 = 18 $\qquad\qquad\qquad$ Answer <u>9</u>
$\qquad\qquad\qquad\qquad$ what times 2 = 18;

$\qquad$ $\frac{7}{21} = \frac{1}{?}$ $\qquad\qquad$ 21 x 1 = 21 $\qquad\qquad\qquad$ Answer <u>3</u>
$\qquad\qquad\qquad\qquad$ what times 7 = 21;

2.7 Problems—Find the missing number:

1. $\dfrac{3}{4} = \dfrac{?}{16}$

2. $\dfrac{5}{6} = \dfrac{?}{12}$

3. $\dfrac{1}{2} = \dfrac{?}{16}$

4. $\dfrac{1}{2} = \dfrac{?}{8}$

5. $\dfrac{2}{?} = \dfrac{10}{50}$

6. $\dfrac{8}{?} = \dfrac{2}{3}$

7. $\dfrac{9}{?} = \dfrac{3}{4}$

8. $\dfrac{10}{?} = \dfrac{5}{7}$

9. $\dfrac{0}{?} = \dfrac{1}{9}$

10. $\dfrac{?}{20} = \dfrac{1}{2}$

11. $\dfrac{?}{15} = \dfrac{2}{5}$

12. $\dfrac{?}{10} = \dfrac{3}{5}$

13. $\dfrac{?}{36} = \dfrac{1}{6}$

14. $\dfrac{?}{30} = \dfrac{1}{10}$

15. $\dfrac{?}{100} = \dfrac{1}{20}$

16. $\dfrac{1}{2} = \dfrac{7}{?}$

17. $\dfrac{4}{6} = \dfrac{8}{?}$

18. $\dfrac{5}{10} = \dfrac{10}{?}$

19. $\dfrac{8}{20} = \dfrac{2}{?}$

20. $\dfrac{9}{15} = \dfrac{3}{?}$

ADDING FRACTIONS WITH LIKE DENOMINATORS

When adding fractions whose denominators (bottom numbers) are the same, just add the numerators (top numbers). The denominator stays the same.

$$\frac{2}{5} + \frac{1}{5} = \frac{3}{5} \qquad \text{or} \qquad \frac{6}{11} + \frac{2}{11} = \frac{8}{11}$$

2.8 Problems—Add the following fractions:

1. $\frac{1}{2} + \frac{1}{2} =$

2. $\frac{1}{3} + \frac{1}{3} =$

3. $\frac{1}{5} + \frac{1}{5} =$

4. $\frac{3}{5} + \frac{1}{5} =$

5. $\frac{2}{6} + \frac{1}{6} =$

6. $\frac{4}{7} + \frac{2}{7} =$

7. $\frac{1}{10} + \frac{7}{10} =$

8. $\frac{3}{12} + \frac{4}{12} =$

9. $\frac{9}{20} + \frac{1}{20} =$

10. $\frac{1}{9} + \frac{6}{9} =$

11. $\begin{array}{r} \frac{7}{11} \\ + \frac{2}{11} \end{array}$

12. $\begin{array}{r} \frac{9}{13} \\ + \frac{4}{13} \end{array}$

13. $\begin{array}{r} \frac{1}{4} \\ + \frac{3}{4} \end{array}$

14. $\begin{array}{r} \frac{1}{15} \\ + \frac{10}{15} \end{array}$

15. $\begin{array}{r} \frac{1}{4} \\ + \frac{3}{4} \end{array}$

16. $\begin{array}{r} \frac{1}{6} \\ + \frac{4}{6} \end{array}$

17. $\begin{array}{r} \frac{2}{10} \\ + \frac{2}{10} \end{array}$

18. $\begin{array}{r} \frac{2}{7} \\ + \frac{1}{7} \end{array}$

19. $\begin{array}{r} \frac{10}{19} \\ + \frac{2}{19} \end{array}$

20. $\begin{array}{r} \frac{21}{25} \\ + \frac{1}{25} \end{array}$

ADDING FRACTIONS WITH UNLIKE DENOMINATORS

You cannot add fractions that do not have the same or "common" denominator (bottom number). To work the problem, first find a common denominator. Then find a new numerator (top number) and as you did fractions with common denominators.

$$\frac{1}{2} + \frac{1}{3}$$

$$\frac{1}{2} + \frac{1}{3} = \frac{?}{6} + \frac{?}{6}$$

$$2/\overline{6} \quad \& \quad 3/\overline{6}$$

3 x 1 = 3 $\frac{3 + 2}{6} = \frac{5}{6}$
2 x 1 = 2

_____

1. Find a common denominator. This will be the lowest number that 2 **and** 3 will divide into without a remainder. 6 is the lowest common denominator in this problem.

2. Find the new numerator by dividing the old denominator into the new one. 2 into 6 and 3 into 6. Multiply the quotient by the old numerator and this gives you the new numerator.

3. Add the numerators and bring the numerator over. Remember, the denominator always remains the same.

4. Reduce if possible.

2.9 Problems—Find the common denominator and add:

1. $\dfrac{1}{2} + \dfrac{1}{4}$

2. $\dfrac{1}{3} + \dfrac{1}{4}$

3. $\dfrac{1}{15} + \dfrac{2}{5}$

4. $\dfrac{3}{4} + \dfrac{1}{5}$

5. $\dfrac{1}{2} + \dfrac{2}{3}$

6. $\dfrac{2}{3} + \dfrac{3}{4}$

7. $\dfrac{1}{6} + \dfrac{2}{7}$

8. $\dfrac{5}{12} + \dfrac{1}{2}$

ADDING UNLIKE DENOMINATORS (CONTINUED)

9. $\underline{3}$
    10

    $\underline{1}$
    + 5

13. $\underline{5}$
     16

     $\underline{3}$
     + 8

10. $\underline{1}$
     4

     $\underline{3}$
     + 10

14. $\underline{1}$
     3

     $\underline{1}$
     + 6

11. $\underline{1}$
     16

     $\underline{1}$
     + 4

15. $\underline{2}$
     3

     $\underline{6}$
     + 10

12. $\underline{2}$
     5

     $\underline{4}$
     + 15

16. $\underline{2}$
     3

     $\underline{1}$
     + 4

17. 5 1
      2
    3 3
    + 4
    ———

19. 2
     9
    10 1
    + 3
    ———

18. 1
     2
    4 2
    + 5
    ———

20. 7 1
      5
      1
    + 3
    ———

SUBTRACTING FRACTIONS WITH LIKE DENOMINATORS

Once you have learned how to add fractions, subtracting becomes easy. All the steps are the same except you subtract the top numbers (numerators) instead of adding:

$$\frac{6}{11} - \frac{2}{11} = \frac{4}{11}$$

2.10 Problems—Subtract the following fractions:

1. $\frac{3}{2} - \frac{2}{2} =$

2. $3 - 1 =$

3. $\frac{2}{3} - \frac{1}{3} =$

4. $\frac{5}{6} - \frac{2}{6} =$

5. $\frac{4}{7} - \frac{1}{7} =$

6. $\frac{9}{10} - \frac{7}{10} =$

7. $\frac{7}{8} - \frac{2}{8} =$

8. $\frac{8}{11} - \frac{3}{11} =$

9. $\frac{11}{20} - \frac{7}{20} =$

10. $\frac{4}{5} - \frac{1}{5} =$

11. $\frac{7}{9}$
    $-\frac{2}{9}$

12. $\frac{10}{12}$
    $-\frac{9}{12}$

13. $\frac{12}{13}$
    $-\frac{9}{13}$

14. $\frac{10}{17}$
    $-\frac{5}{17}$

15. $\frac{93}{100}$
    $-\frac{73}{100}$

16. $\frac{10}{50}$
    $-\frac{9}{50}$

17. $\frac{70}{75}$
    $-\frac{60}{75}$

18. $\frac{20}{30}$
    $-\frac{10}{30}$

19. $\frac{10}{19}$
    $-\frac{3}{19}$

20. $\frac{19}{23}$
    $-\frac{8}{23}$

SUBTRACTING WITH UNLIKE DENOMINATORS

Like addition of fractions, you **cannot work** the problem until the denominators are the same. Use the following steps:

$$\frac{4}{5} - \frac{1}{2}$$

1. Find the lowest number that 5 and 2 will both divide into. The answer is **10.**
2. $\frac{4}{5} - \frac{1}{2} = \frac{x}{10} - \frac{x}{10}$ Find the new numerators by dividing the old denominators (5 & 2) into the new one (10) and multiply this by the old numbers (4 & 1).
3. $\frac{4}{5} - \frac{1}{2} = \frac{8}{10} - \frac{5}{10} = \frac{3}{10}$

Remember, in adding and subtracting fractions, the top number only is added or subtracted. The bottom number stays the same.

2.11 Problems—Subtract the following fractions:

1. $\frac{1}{2} - \frac{1}{3} =$

2. $\frac{3}{4} - \frac{1}{5} =$

3. $\frac{5}{6} - \frac{1}{3} =$

4. $\frac{9}{10} - \frac{2}{5} =$

5. $\frac{1}{2} - \frac{1}{6} =$

6. $\frac{6}{7} - \frac{2}{3} =$

7. $\frac{3}{4} - \frac{2}{8} =$

8. $\frac{5}{6} - \frac{1}{12} =$

9. $\frac{4}{5}$
   $- \frac{2}{3}$

10. $\frac{8}{10}$
    $- \frac{1}{5}$

11. $\dfrac{4}{6}$
   $-\dfrac{1}{3}$

16. $\dfrac{6}{7}$
   $-\dfrac{3}{5}$

12. $\dfrac{1}{2}$
   $-\dfrac{1}{4}$

17. $\dfrac{7}{10}$
   $-\dfrac{3}{20}$

13. $\dfrac{4}{6}$
   $-\dfrac{1}{2}$

18. $\dfrac{43}{50}$
   $-\dfrac{11}{100}$

14. $\dfrac{7}{8}$
   $-\dfrac{3}{16}$

19. $\dfrac{6}{7}$
   $-\dfrac{3}{10}$

15. $\dfrac{2}{5}$
   $-\dfrac{1}{6}$

20. $\dfrac{1}{8}$
   $-\dfrac{1}{9}$

SUBTRACTING MIXED NUMBERS

To subtract mixed numbers, you may first subtract the whole number, then subtract the fractions:

$$7\frac{3}{5}$$
$$-\,3\frac{1}{5}$$
$$4\frac{2}{5}$$

Subtract 3 from 7 = 4
Subtract $\frac{1}{5}$ from $\frac{3}{5}$ = $\frac{2}{5}$

$$14\frac{7}{8} = 14\frac{21}{24}$$
$$-\,3\frac{5}{6} = 3\frac{20}{24}$$
$$11\frac{1}{24}$$

To subtract mixed numbers with unlike denominators, you must first find a common denominator. Then subtract as above.

To subtract a fraction from a whole number, change the whole number to a mixed number. See the example to the left.

$$7 = 6\frac{3}{3}$$
$$-\,\frac{1}{3} = \frac{1}{3}$$
$$6\frac{2}{3}$$

Problems—Subtract the following mixed numbers:

2.12 Problems

1. $7\frac{2}{3}$
    $-1\frac{1}{3}$
    $\frac{3}{}$

2. $6\frac{4}{5}$
    $-2\frac{1}{5}$
    $\frac{5}{}$

3. $10\frac{3}{4}$
    $-5\frac{1}{4}$
    $\frac{4}{}$

4. $4\frac{3}{2}$
    $-1\frac{1}{2}$
    $\frac{2}{}$

5. $7\frac{5}{6}$
    $-6\frac{3}{6}$
    $\frac{6}{}$

6. $9\frac{6}{7}$
    $-2\frac{3}{7}$
    $\frac{7}{}$

7. $11\frac{9}{10}$
    $-1\frac{7}{10}$
    $\frac{10}{}$

8. $17\frac{19}{20}$
    $-6\frac{11}{20}$
    $\frac{20}{}$

9. $2\frac{40}{100}$
    $-1\frac{35}{100}$
    $\frac{100}{}$

10. $3\frac{10}{11}$
    $-2\frac{3}{11}$
    $\frac{11}{}$

11. $5\frac{11}{25}$
    $-1\frac{9}{25}$
    $\frac{25}{}$

12. $6\frac{49}{50}$
    $-2\frac{48}{50}$
    $\frac{50}{}$

13. $7\frac{10}{17}$
    $-5\frac{5}{17}$
    $\frac{17}{}$

14. $12\frac{2}{3}$
    $-10\frac{1}{3}$
    $\frac{3}{}$

15. $6\frac{4}{5}$
    $-5\frac{3}{5}$
    $\frac{5}{}$

16. 8 $\underline{5}$
        6
     - 3 $\underline{3}$
     ___ $\underline{6}$

17. 9 $\underline{4}$
        7
     - 6 $\underline{2}$
     ___ $\underline{7}$

18. 18 $\underline{6}$
        8
     - 12 $\underline{3}$
     ___ $\underline{8}$

19. 22 $\underline{2}$
        3
     - 19 $\underline{1}$
     ___ $\underline{2}$

20. 100 $\underline{4}$
        5
     - 93 $\underline{2}$
     ___ $\underline{10}$

MULTIPLYING FRACTIONS

To multiply fractions, multiply the top numbers together **and** multiply the bottom numbers together. Examples below:

$$\frac{2}{3} \times \frac{3}{5} = \frac{6}{15}$$
$$\frac{1}{2} \times \frac{1}{2} = \frac{1}{4}$$
$$\frac{3}{4} \times 5 = \frac{3}{4} \times \frac{5}{1} = \frac{15}{4}$$

$$7\frac{2}{3} \times \frac{1}{2} = \frac{23}{3} \times \frac{1}{2} = \frac{23}{6}$$ Change the mixed number to a fraction.

2.13 Problems—Multiply the following fractions:

1. $\frac{1}{2} \times \frac{1}{2} =$

6. $\frac{2}{3} \times \frac{2}{3} =$

2. $\frac{1}{3} \times \frac{1}{4} =$

7. $\frac{4}{6} \times \frac{5}{6} =$

3. $\frac{2}{6} \times \frac{1}{7} =$

8. $\frac{9}{10} \times \frac{2}{10} =$

4. $\frac{3}{9} \times \frac{2}{2} =$

9. $\frac{1}{11} \times \frac{3}{9} =$

5. $\frac{1}{4} \times \frac{1}{3} =$

10. $\frac{1}{5} \times \frac{4}{6} =$

11. $\dfrac{1}{3} \times \dfrac{1}{2} =$

12. $\dfrac{1}{5} \times \dfrac{2}{4} =$

14. $\dfrac{2}{7} \times \dfrac{3}{7} =$

15. $\dfrac{7}{10} \times \dfrac{1}{10} =$

16. $\dfrac{8}{9} \times \dfrac{1}{3} =$

17. $\dfrac{3}{8} \times \dfrac{1}{2} =$

18. $\dfrac{6}{7} \times \dfrac{3}{4} =$

19. $\dfrac{1}{2} \times \dfrac{3}{2} =$

20. $\dfrac{1}{3} \times \dfrac{6}{7} =$

DIVIDING FRACTIONS

It is easier in mathematics to change all division problems dealing with fractions to multiplication. This can be done easily by inverting, (turn upside down) the second fraction only. Then just multiply.

Examples:

$$\frac{1}{2} \div \frac{1}{3} = \frac{1}{2} \times \frac{3}{1} = \frac{3}{2} = 1\frac{1}{2}$$

$$\frac{3}{4} \div 6 = \frac{3}{4} \times \frac{1}{6} = \frac{3}{24} = \frac{1}{8}$$

$$7 \div \frac{2}{3} = \frac{7}{1} \times \frac{3}{2} = \frac{21}{2} = 10\frac{1}{2}$$

Remember that a whole number may be placed over 1. If the division problem contains mixed numbers, simply change them to improper fractions.

Examples:

$$3\frac{2}{5} \div 2\frac{1}{2} = \frac{17}{5} \div \frac{5}{2} = \frac{17}{5} \times \frac{2}{5} = \frac{34}{25} = 1\frac{9}{25}$$

Always remember to change **only** the second fraction upside down. The first fraction or number stays the same.

2.14 Problems—Divide the following fractions:

1. $\frac{7}{9} \div 2 =$

2. $\frac{3}{4} \div 3 =$

3. $\frac{3}{5} \div 4 =$

4. $\frac{7}{8} \div 10 =$

5. $\frac{9}{10} \div 5 =$

6. $\frac{1}{2} \div 6 =$

7. $\frac{1}{3} \div 7 =$

8. $\frac{1}{2} \div \frac{1}{2} =$

9. $\dfrac{1}{3} \div \dfrac{1}{3} =$

10. $7 \div \dfrac{1}{2} =$

11. $5 \div \dfrac{1}{2} =$

12. $10 \div \dfrac{1}{10} =$

13. $2 \div \dfrac{1}{6} =$

14. $4 \div \dfrac{1}{7} =$

15. $6 \div \dfrac{2}{5} =$

16. $\dfrac{1}{4} \div \dfrac{1}{8} =$

17. $\dfrac{1}{8} \div \dfrac{1}{4} =$

18. $\dfrac{2}{3} \div \dfrac{1}{4} =$

19. $\dfrac{6}{11} \div \dfrac{1}{2} =$

20. $\dfrac{7}{9} \div \dfrac{1}{3} =$

EXPRESSING REMAINDERS AS A FRACTION

When dividing and the quotient has a remainder, it is better to express this remainder as a fraction. Put the number that is "left over" as the numerator (top Number) and the divisor will be the denominator (bottom number). Look at the examples below.

```
    3 1                 1 9                 9 2
      5                  10                  3
  ─────              ─────              ─────
5/ 16             10/ 19              3/ 29
   15                  10                  27
 ────              ────               ────
    1                   9                   2
```

2.15 Problems—Divide and express the remainders as fractions. Reduce the fractions to the lowest terms:

1. 2⟌9

2. 4⟌15

3. 10⟌22

4. 5⟌62

5. 6⟌63

6. 7⟌50

7. 8⟌19

8. 11⟌100

9. 12⟌14

10. 15⟌18

2.16 Problems—Divide (Some have remainders)

1. 9/ 81‾

2. 8/ 192‾

    4/ 800‾

3. 5/ 525‾

4. 24/ 1248‾

5. 24/ 504‾

6. 36/ 1476‾

7. 10/ 370‾

9. 18/ 3600‾

10. 11/ 121‾

11. 12/ 144‾

12. 42/ 126‾

13. 10/ 1050‾

14. 17/ 85‾

15. 6/ 37‾

16. 10/ 102‾

TEST—CHAPTER II—FRACTIONS

Add:

1. $\dfrac{3}{5} + \dfrac{1}{5} =$

2. $\dfrac{1}{3} + \dfrac{1}{3} =$

3. $\dfrac{4}{7} + \dfrac{2}{7} =$

4. $\begin{array}{r} \dfrac{1}{3} \\ \dfrac{2}{5} \\ + \dfrac{}{} \end{array}$

5. $\begin{array}{r} \dfrac{1}{2} \\ \dfrac{3}{4} \\ + \dfrac{}{} \end{array}$

Subtract:

6. $\dfrac{3}{6} - \dfrac{2}{6} =$

7. $\begin{array}{r} \dfrac{3}{5} \\ - \dfrac{1}{5} \end{array}$

8. $\begin{array}{r} \dfrac{7}{9} \\ - \dfrac{3}{9} \end{array}$

9. $\begin{array}{r} \dfrac{7}{8} \\ - \dfrac{1}{8} \end{array}$

10. $\begin{array}{r} \dfrac{1}{2} \\ - \dfrac{1}{3} \end{array}$

Multiply:

11. $\dfrac{1}{2} \times \dfrac{1}{2} =$

12. $\dfrac{3}{4} \times \dfrac{7}{8} =$

13. $3 \times \dfrac{1}{4} =$

14. $6\dfrac{1}{2} \times \dfrac{1}{4} =$

15. $7\dfrac{3}{4} \times 3\dfrac{2}{3} =$

Divide:

16. $\dfrac{1}{3} \div \dfrac{1}{2} =$

17. $6 \div \dfrac{1}{5} =$

18. $\dfrac{2}{3} \div 7 =$

19. $\dfrac{1}{4} \div \dfrac{3}{4} =$

20. $7\dfrac{1}{2} \div \dfrac{3}{4} =$

Reduce:

21. $\dfrac{10}{15} =$

22. $\dfrac{4}{8} =$

23. $\dfrac{15}{25} =$

24. $\dfrac{18}{2} =$

25. $\dfrac{17}{3} =$

CHAPTER 2 ANSWERS

Chapter 2.1

1	2
2	9

Chapter 2.2

1

1	21	36	55	74	90
4	22	38	56	75	92
6	24	39	58	76	94
8	25	40	60	77	95
9	26	42	62	78	96
10	27	44	64	80	98
12	28	45	65	81	99
14	30	26	66	82	100
15	32	47	68	84	
16	33	50	69	85	
18	34	52	70	86	
20	35	54	72	88	

Chapter 2.3

1	=		11	=
2	=		12	≠
3	≠		13	≠
4	≠		14	≠
5	≠		15	=
6	≠		16	≠
7	=		17	=
8	=		18	≠
9	=		19	≠
10	=		20	≠

Chapter 2.4

1	2/3	11	3/4
2	1/2	12	3/5
3	2/3	13	1/4
4	1/3	14	1/3
5	1/2	15	3/4
6	5/6	16	1/3
7	1/5	17	2/3
8	1/5	18	1/5
9	1/2	19	1/10
10	2/3	20	7/10

Chapter 2.5

1	6	6	5 2/9
2	5	7	3
3	6 2/5	8	2 1/2
4	3 1/2	9	1 1/2
5	2 1/4	10	3 1/2

Chapter 2.6

1	3/2	11	24/7
2	5/2	12	27/5
3	19/74	13	14/3
4	91/9	14	7/2
5	7/2	15	5/2
6	17/3	16	5/3
7	52/7	17	39/5
8	21/2	18	67/10
9	34/3	19	11/6
10	14/6	20	11/4

Chapter 2.7

1	12	11	6
2	10	12	6
3	8	13	6
4	4	14	3
5	10	15	5
6	12	16	14
7	12	17	12
8	14	18	20
9	90	19	5
10	10	20	5

Chapter 2.8

1	1	11	9/11
2	2/3	12	1
3	2/5	13	1
4	4/5	14	11/15
5	1/2	15	1
6	6/7	16	5/6
7	4/5	17	2/5
8	7/12	18	3/7
9	1/2	19	12/19
10	7/9	20	22/25

Chapter 2.9

1	4/8 + 2/8 = 6/8 = 3/4	11	5/16
2	7/12	12	2/3
3	1/15 + 6/15 = 7/15	13	11/16
4	19/20	14	1/2
5	3/6 + 4/6 = 7/6 = 1 1/16	15	1 4/15
6	1 5/12	16	1 1/6
7	7/42 + 12/42 = 19/42	17	9 1/4
8	11/12	18	4 7/10
9	1/2	19	10 5/9
10	11/20	20	7 1/3

Chapter 2.10

1	1/2	11	5/9
2	2	12	1/12
3	1/3	13	3/13
4	1/2	14	5/17
5	3/7	15	1/5
6	1/5	16	1/50
7	5/8	17	2/15
8	5/11	18	1/3
9	1/5	19	7/19
10	3/5	20	11/23

Chapter 2.11

1	1/6	11	1/3
2	11/20	12	1/4
3	1/2	13	1/6
4	1/2	14	11/16
5	1/3	15	7/30
6	4/21	16	9/35
7	1/2	17	11/20
8	3/4	18	3/4
9	2/15	19	39/70
10	3/5	20	1/72

Chapter 2.12

1	6 1/3		11	4 2/25
2	4 3/5		12	4 1/50
3	5 1/2		13	2 5/17
4	4		14	2 1/3
5	1 1/3		15	1 1/5
6	7 3/7		16	5 1/3
7	10 1/5		17	3 2/7
8	11 2/5		18	6 3/8
9	1 1/20		19	2 1/6
10	1 7/11		20	7 1/10

Chapter 2.13

1	1/4		11	1/6
2	1/12		12	1/10
3	1/21		13	1/4
4	1/3		14	6/49
5	1/12		15	4/57
6	4/9		16	8/27
7	5/9		17	3/16
8	9/20		18	9/14
9	1/33		19	3/4
10	2/15		20	2/7

Chapter 2.14

1	7/18		11	10
2	1/4		12	100
3	3/20		13	12
4	7/80		14	28
5	9/50		15	15
6	1/12		16	2
7	1/21		17	1/2
8	1		18	2 2/3
9	1		19	1 1/11
10	14		20	2 1/3

Chapter 2.15

1	4 1/2		6	49 1/7
2	3 3/4		7	2
3	2 1/5		8	9 1/11
4	12 2/5		9	1 1/6
5	10 1/2		10	1 1/5

Chapter 2.16

1	9	9	200
2	24	10	11
3	200	11	12
4	105	12	3
5	52	13	3
6	21	14	5
7	41	15	5
8	37	16	10 1/5

TEST

1	4/5	14	1 5/8
2	2/3	15	28 5/12
3	6/7	16	2/3
4	11/15	17	30
5	1 1/4	18	2/21
6	1/6	19	1/3
7	2/5	20	10
8	4/9	21	2/3
9	3/4	22	1/2
10	1/6	23	3/5
11	1/4	24	9
12	21/32	25	5 2/3
13	3/4		

CHAPTER 3

DECIMALS

Every number has a decimal point. In some numbers you can see it while in others you cannot actually see it. Remember, if you don't see a decimal point, it is understood to be behind or to the right of the last digit:

$$17 = 17.$$
$$2 = 2.$$
$$135 = 135.$$

We have already studied numbers and their place value on the left of the decimal point. Now let's learn the values to the right of the decimal point. Below is a table of values that you must learn to achieve a good understanding of decimals.

Decimal Point	Tenths	Hundredths	Thousandths	Ten Thousandths	Hundred Thousandths	Millionths
.	5	1	9	2	4	5

The first number to the **left** of the decimal point is called Ones, then Tens, etc. But the first number to the **right** of the decimal point is called tenths, then hundredths. Both sides of the decimal increase by multiples of 10. The difference is that in decimals you start with the highest value (tenths) and the further you go to the right, the **smaller** the value of the number.

To read decimals, never say the word "point", say "and" instead. Whatever place the last digit is in determines how the number is read. For example, 21.621 would read: "twenty one **and** six hundred twenty one thousandths", since the last digit (1) is in the thousandths place. **Remember, a number has only one decimal point.**

DECIMALS (CONTINUED)

Examples:	115.6	"One hundred fifteen **and** six tenths"
	7.37	"Seven **and** thirty seven hundredths"
	.45	"Forty five hundredths"

3.1 Problems—Write in words the following numbers:

1. .6

2. 3.1

3. 42.3

4. .67

5. 1.12

6. 1.992

7. 7.02

8. 53.42

9. 9.901

10. 10.101

3.2 Write the following in numbers:

1. Seven-tenths   _____

2. Three and four-tenths   _____

3. Sixty one and three-tenths   _____

4. Seventeen and two-hundredths   _____

5. Thirty-two hundredths   _____

6. Nine and nine-hundredths   _____

7. Four and six-thousandths   _____

8. Fifty-two hundredths   _____

9. Eight and six-tenths   _____

10. Ninety three and seven-hundredths   _____

ADDITION OF DECIMALS

When adding decimals, keep the decimal point in a straight line going down and add, as you would whole numbers. If a problem is written across, rewrite it up and down.

$.8 + 7.3 + .09 =$

$$
\begin{array}{r}
.8 \\
7.3 \\
+\ \ .09 \\
\hline
8.19
\end{array}
$$

$$
\begin{array}{r}
.1 \\
2.37 \\
11.204 \\
+\ \ .9076 \\
\hline
14.5816
\end{array}
$$

3.3 Problems—Add:

1. $.6 + 6.6 + .06 =$

2. $7.7 + .02 + 3.1 =$

3. $3.11 + .256 =$

4. $7.67 + .7 =$

5. $88 + 8.8 + .8 =$

6. $.105 + 2.051 =$

7. $34.56 + 65.43 =$

8. $62.19 + 72.03 =$

9. $.91 + .70 + .62 =$

10. $89 + 4 + 602 =$

11. $4.4 + .44 + 4 =$

12. .012
    1.71
+ 9.32

17. 9.25
    .274
+ .43

13. 92.73
    .37
+ .111

18. .61
    6.6
+ 8.3

14. 111.1
    1.1
+ 1.011

19. 71.9
    .925
+ 7.

15. 73.4
  12.49
+ .712

20. 141.5
    9.46
+ .73

16. 82.28
+ 111.73

SUBTRACTION OF DECIMALS

The rule for subtracting decimals is the same as adding decimals, line up the decimal point vertically. However, it is necessary to place zeros when one of the numbers does not have the same number of digits:

$$
\begin{array}{cccc}
1.75 & = & 8.750 \\
-\ \underline{1.326} & & \underline{1.326}
\end{array}
$$

There are imaginary zeros after the 8.75, and by putting a zero over the 1.326 you are less likely to make the mistake of bringing down the six instead of subtracting 6 from 0 (10), which is **4.**

3.4 Problems—Subtract and check:

1. .9 - .7 =

2. 1.1 - .9 =

3. 12.56 - 10.11 =

4. 17.60 - 13.50 =

5. 2.25 - 1.23 =

6. 62.6 - .5 =

7. 1.15 - .15 =

9. 8.51 - 7.12 =

8. .75 - .65 =

10. 14.1 - 1.9 =

SUBTRACTION OF DECIMALS (CONTINUED)

11. 13.6
   - 7.6

14. 96.87
   - 14

12. 14.54
   - .621

15. 7.08
   - .03

13. 19.02
   - 19.01

16. 7.7
   - 1.8

17. 9.7
 - 1.34

19. 79.37
 - 79.31

18. 100.02
 - 90.02

20. 43.72
 - .732

MULTIPLICATION OF DECIMALS

To multiply decimals, you follow the same steps as you did in multiplying whole numbers. When you find your answer you point off from right to left the decimal point. The number of places depends on the total number of numerals that are to the right of the decimal point in the two numbers you have multiplied.

$$
\begin{array}{r}
1.7 \\
\times\ \ .2 \\
\hline
.34
\end{array}
$$

←

There are 2 numbers (total) to the right of the decimal point.

Move the decimal point 2 places to the left in your answer.

Always begin at the far right. Add zeros if necessary.

1.2 (1)	.12 (2)	.06 (2)	1.1 (1)
x .3 (1)	x .12 (2)	x .03 (2)	x .22 (2)
.36 (2)	24	18	22
	12	00	22
	.0144 (4)	.0018 (4)	.242 (3)

3.5 Problems—Multiply:

1. 3.1
    x 1.2

2. .31
    x 1.2

3. .31
    x .12

4. .43
    x 1.4

5. .43
    x 14

6. .007
    x 2

MULTIPLICATION OF DECIMALS (CONTINUED)

7. .03
    x 03

12. 291
    x .2

17. 13.6
    x .2

8. 4
    x .5

13. 41.02
    x .3

18. 5
    x .5

9. 71
    x .09

14. .5
    x 6

19. 8.1
    x 11

10. .12
    x .09

15. .5
    x 1

20. .5
    x .8

11. 62
    x .3

16. 1.61
    x .2

DIVISION OF DECIMALS

1. Dividing when there is a decimal point in the dividend:

    $5\overline{)1.5}$ $6\overline{).18}$ $2\overline{)44.4}$

    Move the decimal point straight up into the answer (quotient) and then divide as you would whole numbers:

.3	.03	22.2
$5\overline{)1.5}$	$6\overline{).18}$	$2\overline{)44.4}$
15	0	4
0	18	04
	18	04
	0	0

3.6 Divide the following:

1. $2\overline{).14}$

2. $4\overline{).8}$

3. $6\overline{)1.2}$

4. $3\overline{)9.9}$

5. $5\overline{)1.55}$

6. $7\overline{).7}$

7. $9\overline{)9.9}$

8. $10\overline{)15.0}$

9. $3\overline{)6.3}$

10. $2\overline{).48}$

DIVISION OF DECIMALS (CONTINUED)

11. $7\overline{)7.7}$

16. $7\overline{)14.7}$

12. $5\overline{)2.5}$

17. $3\overline{).93}$

13. $5\overline{).35}$

18. $2\overline{).64}$

14. $4\overline{).88}$

19. $4\overline{)8.0}$

15. $8\overline{)8.8}$

20. $6\overline{)6.6}$

DIVISION OF DECIMALS (CONTINUED)

2. Dividing when there is a decimal in the divisor:

$$.5\overline{)\,15\,} \qquad .6\overline{)\,18\,} \qquad .02\overline{)\,444\,}$$

Move the decimal to the right the number of places necessary to make the decimal come after the last numeral.

MOVE THE DECIMAL IN THE DIVIDEND THE SAME NUMBER OF PLACES:

$$5.\overline{)\,15.0\,} \qquad 6.\overline{)\,180\,} \qquad 02.\overline{)\,44400\,}$$

Remember: All numbers that do not show a decimal point have one at the far right. Add as many zeros as necessary to move the decimal the exact number of places in the dividend as in the divisor.

After you have moved the decimal in the divisor and the dividend, place the decimal point straight up into the quotient as you did in the problems and examples on the previous page; Then divide:

$$.5\overline{)\,15.0\,}^{\,.} \qquad .6\overline{)\,18.0\,}^{\,.} \qquad .02\overline{)\,444.00\,}^{\,.}$$

3. Dividing when there is a decimal point shown in the divisor and the dividend.

In mathematics you have to move the decimal point in the divisor behind the number. The reason is to make it a whole number by multiplying by 10, 100, 1000, etc. Always move the decimal point the same number of places in both the divisor and the dividend. The decimal point in the divisor has to end up behind the last digit. The decimal point in the dividend simply ends up as many places as it had to be moved.

$$.2\overline{)\,1.4\,} \qquad .7\overline{)\,.14\,} \qquad .07\overline{)\,.147\,} \qquad .007\overline{)\,14.700\,}$$

3.7 Problems—Divide:

1. $.5\overline{)15}$

4. $.3\overline{)9}$

7. $.7\overline{)147}$

2. $5\overline{)1.5}$

5. $.03\overline{)9}$

8. $7\overline{)147}$

3. $.5\overline{).15}$

6. $3\overline{).9}$

9. $8\overline{)6.4}$

10. $8 \overline{).64}$

14. $.6 \overline{).36}$

18. $.2 \overline{).42}$

11. $.04 \overline{)1.6}$

15. $6 \overline{).36}$

19. $.10 \overline{)100}$

12. $.4 \overline{).16}$

16. $.2 \overline{)6}$

20. $.5 \overline{)52.5}$

13. $.5 \overline{)2.5}$

17. $2 \overline{)6.2}$

TEST—CHAPTER III—DECIMALS

Add:

1. .2 + 2 + 2.2 =

2. .11 + 10.1 + 1 =

3. 2.3
    1.47
    + 11.111

4. 5.5
    5.531
    + .112

5. 6.7
    1.73
    + 7.714

Subtract:

6. 7.96 - 3.33 =

7. 8.93
    - 1.23

8. 7.39
    - .464

Multiply:

9. .07
    x   3

10. 1.11
    x   .2

11. .143
    x .1

12. .009
    x .8

13. 34.5
    x 2

14. .711
    x .04

15. 1298
    x   .3

<u>TEST—CHAPTER III—DECIMALS</u>

16. $.6\overline{)\,12}$ 17. $6\overline{)\,1.2}$ 18. $.6\overline{)\,.12}$

19. $.06\overline{)\,1.2}$ 20. $.6\overline{)\,1.20}$

ANSWERS CHAPTER 3

Chapter 3.1

1 Six Tenths

2 Three and One Tenth

3 Forty-Two and Three Hundreths

4 Sixty-Seven Hundreths

5 One and Twelve Hundreths

6 One and Nine Hundred Ninety-Two Thousandths

7 Seven and Two Hundreths

8 Fifthy-Three and Fourty-Two Hundreths

9 Nine and Nine Hundred One Thousandths

10 Ten and One Hundred One Thousandths

Chapter 3.2

1	0.7		6	9.09
2	3.4		7	4.006
3	61.3		8	0.52
4	17.02		9	8.6
5	0.32		10	93.07

Chapter 3.3

1	7.26		11	8.84
2	10.82		12	11.042
3	3.366		13	93.211
4	8.37		14	113.211
5	97.6		15	86.602
6	2.156		16	194.01
7	99.99		17	9.954
8	134.22		18	15.51
9	2.23		19	79.825
10	695		20	151.69

Chapter 3.4

1	0.2	11	6
2	0.2	12	13.919
3	2.45	13	0.01
4	4.1	14	82.87
5	1.02	15	7.05
6	62.1	16	5.9
7	1	17	8.36
8	0.10	18	10
9	1.39	19	0.06
10	12.2	20	42.988

Chapter 3.5

1	3.72	11	18.6
2	0.372	12	58.2
3	0.0372	13	123.06
4	0.602	14	3
5	6.02	15	0.5
6	0.014	16	0.322
7	0.0009	17	40.8
8	2	18	2.5
9	6.39	19	89.1
10	0.0108	20	0.4

Chapter 3.6

1	0.07	11	1.1
2	0.125	12	0.5
3	0.2	13	0.007
4	3.3	14	0.22
5	0.31	15	1.1
6	0.1	16	2.1
7	1.1	17	0.31
8	1.5	18	0.32
9	2.1	19	2
10	0.24	20	1.1

Chapter 3.7

1	30	8	21
2	0.3	9	0.8
3	0.3	10	0.08
4	30	11	40
5	300	12	0.4
6	0.3	13	5
7	210	14	60

15	0.06		18	2.1
16	30		19	1000
17	3.1		20	105

TEST Chapter 3

1	4.4		11	0.0143
2	11.21		12	0.0081
3	148.81		13	69
4	11.143		14	0.0244
5	1.144		15	389.4
6	4.63		16	20
7	7.7		17	0.2
8	6.926		18	0.2
9	0.21		19	20
10	0.222		20	2

CHAPTER 4

PERCENT:

The word "percent" means the same as "hundredths" (2 decimal places). Sometimes it is easier to use percent rather than decimals or fractions. Numerous examples in sports, sales discounts, and business can be expressed as a percentage "Rick Berry shot 90% from the foul line". "Toyota sells for 20% off the regular price". "General Motors increased sales by 10% over last year". Percents, fractions and decimals are all "related" to each other. We have studied fractions and decimals, now we shall study percent.

THE SYMBOL FOR PERCENT IS %.

Problems:

1. Find 5 examples of % on the internet, sales flier or magazine.

2. Explain why it is sometimes easier to use % rather than fractions.

CHANGING A PERCENT TO A DECIMAL:

To change a percent to a decimal, move the decimal point 2 places to the left and drop the % sign. Remember that if you do not see a decimal point, it is behind, or to the right of, the last number!

Examples: 20% = .20   10% = .10   95% = .95   105% = 1.05

          1% = .01   5% = .05   .5% = .005   200% = 2.00

4.1 Problems—Change the following Percents to Decimals:

1. 17% =

2. 50% =

3. 75% =

4. 25% =

5. 2% =

6. 9% =

7. 100% =

8. 200% =

9. 111% =

10. 1.5% =

12. 2% =

13. 22% =

14. .6% =

15. .04% =

16. 700% =

17. 1000% =

18. 14% =

19. 11.5% =

20. 22.67% =

CHANGING A DECIMAL TO A PERCENT:

To change a decimal to a percent, move the decimal point 2 places to the right and add the percent sign (%). <u>Most</u> of the time the decimal point will end up behind the number, <u>but not always.</u> The rule is very clear—move the decimal point 2 places to the right, and add the percent sign (%).

Examples: .20 = 20%   .15 = 15%   .70 = 70%   .165 = 16.5%

              1 = 100%   2.5 = 250%   .04 = 4%

4.2 Problems—Change the following Decimals to Percent:

1. .10 =   11. .01 =

2. .20 =   12. .11 =

3. .15 =   13. .75 =

4. .30 =   14. 1.00 =

5. .45 =   15. 2 =

6. .62 =   16. 1.50 =

7. .50 =   17. 3 =

8. .74 =   18. .625 =

9. .09 =   19. .375 =

10. .05 =   20. .01 =

CHANGING A FRACTION TO A PERCENT:

To change a fraction to a percent, change the fraction to a decimal, <u>then</u> change the decimal to a percent.

Example: Change ½ to a percent.

```
      .5        Divide 2 into 1 to find the decimal.
   2/ 1.0
      1 0
        0
```

.5 = 50%   Change .5 to a percent by moving the decimal point two places to the right and then put the % sign. Add zero's if necessary.

4.3 Problems—Change the following fractions to a decimal and then a percent:

1. $\dfrac{1}{4} = $ $\quad \dfrac{.25}{4/\overline{1.00}} = 25\%$

2. $\dfrac{3}{4}$

3. $\dfrac{1}{5}$

4. $\dfrac{2}{5}$

5. $\dfrac{3}{5}$

6. $\dfrac{1}{10}$

7. $\dfrac{3}{10}$

8. $\dfrac{1}{8}$

9. $\dfrac{3}{8}$

10. $\dfrac{5}{8}$

11. $\dfrac{1}{7}$

12. $\dfrac{9}{20}$

13. $\dfrac{1}{50}$

14. $\dfrac{1}{2}$

15. $\dfrac{9}{10}$

16. $\dfrac{7}{10}$

17. $\dfrac{7}{20}$

18. $\dfrac{3}{20}$

19. $\dfrac{1}{20}$

20. $\dfrac{1}{3}$

FINDIND PERCENT OF A NUMBER:

To find the percent of a number, just follow these easy steps:

Find 30% of 60.   **(Hint: "of" means times)**

30% = .30   Change the 30% to a decimal by dropping the % sign, and moving the decimal point 2 places
to the left.

.30 of 60 = .30 x 60   Change the word "of" to a multiplication sign, and multiply.

```
   .30
 x 60
 ‾‾‾‾
   00
 180
‾‾‾‾‾
18.00
```
                    18 is your answer. It is a number that represents 30% of 60.

Examples: 10% of 50 = .10 x 50 = 5
01% of 200 = .01 x 200 = 2
100% of 5 = 1.00 x 5 = 5

4.4 Problems—Find the % of the following:

1. 10% of 30 =

2. 15% of 100 =

3. 5% of 50 =

4. 1% of 30 =

5. 20% of 60 =

6. 25% of 100 =

7. 30% of 68 =

8. 35% of 200 =

9. 50% of 6 =

10. 75% of 400 =

11. 100% of 9 =

12. 200% of 12 =

13. 300% of 3 =

14. 2% of 50 =

15. 40% of 60 =

16. 60% of 40 =

17. 70% of 90 =

18. 80% of 30 =

19. 85% of 30 =

20. 90% of 200 =

RATIO:

The relationship between two numbers when compared by dividing is called the **ratio.**

The ratio of boys to girls in your class is a good example. You could also compare the boys to the total number of people in the class. If there are 15 boys and 20 girls in a class, then the following ratios would be used.

Boys to girls = 15 to 20 or 15 : 20
Girls to boys = 20 to 15 or 20 : 15
Boys to total = 15 to 35 or 15 : 35
Girls to total = 20 to 35 or 20 : 35

The word "to" can be substituted for a colon (:). If it is necessary to express a ratio as a fraction or as a division problem, it may be done in the following manner:

$$1 : 3 \qquad \frac{1}{3} \qquad 3\overline{/\ 1}$$

PROPORTION:

When 2 ratios are compared to each other and they are equal, they are said to form a **proportion**. For example, $3 : 4 = 6 : 8$ is a proportion $\frac{3}{4} = \frac{6}{8}$ To determine if 2 ratios are a proportion, multiply the two outside numbers together; if they are equal, then multiply the two inside numbers and if they are also equal the 2 ratios are a proportion.

$$3 : 4 = 6 : 8 \qquad 3 \times 8 = 24$$
$$4 \times 6 = 24$$

If the two ratios are written like fractions you can cross—multiply they are proportion.

If the two products are the same,

Examples	$\frac{6}{18} = \frac{2}{6}$	$6 \times 6 =$	36
		$18 \times 2 =$	36

$$\frac{6}{9} \neq \frac{1}{10} \qquad 6 \times 10 = 60$$
$$9 \times 1 = 9$$

4.5 Problems—Place = or ≠

1. $\dfrac{1}{2}$ __ $\dfrac{4}{8}$

2. $\dfrac{3}{5}$ __ $\dfrac{6}{10}$

3. $\dfrac{7}{9}$ __ $\dfrac{3}{10}$

4. $\dfrac{4}{5}$ __ $\dfrac{1}{2}$

5. $\dfrac{6}{7}$ __ $\dfrac{10}{70}$

6. $\dfrac{1}{3}$ __ $\dfrac{4}{5}$

7. 6 : 7 __ 1 : 2

8. 4 : 5 __ 10 : 50

9. 1 : 2 __ 3 : 6

10. 4 : 10 __ 5 : 6

11. $\dfrac{1}{6}$ __ $\dfrac{2}{3}$

12. $\dfrac{4}{12}$ __ $\dfrac{1}{3}$

13. $\dfrac{6}{8}$ __ $\dfrac{3}{4}$

14. 5 : 10 __ 10 : 20

15. 7 : 11 __ 11 : 7

16. 1 : 4 __ 4 : 16

17. 3 : 4 __ 75 : 100

18. 5 : 15 __ 1 : 3

19. 10 : 20 __ 1 : 2

20. 7 : 10 __ 10 : 70

4.6 Problems—Fill in the missing numbers:

Fraction	Ratio	Percent
1/2		
	1 : 4	
		75%
1/5		
2/5		
	1 : 10	
	3 : 10	
	3 : 5	
		100%
4/10		
		200%
	6 : 10	
	3 : 50	
9/20		
	11 : 20	
	20 : 20	
	10 : 50	
19/50		
	1 :100	
5/100		
		10%

<u>TEST—CHAPTER IV—PERCENT—RATIO</u>

<u>Change to %:</u>

1. .15

4. 1.00

2. .73

5. .05

3. .21

<u>Change to Decimals:</u>

6. 14%

9. 250%

7. 50%

10. 1%

8. 20%

<u>Find % of the Number:</u>

11. 50% of 60 =

14. 10% of 100 =

12. 20% of 50 =

15. 100% of 20 =

13. 25% of 100 =

Change to Fractions:

16. 10% =

17. 15% =

18. 35% =

19. 50% =

20. 75% =

ANSWERS CHAPTER 4

Chapter 4.1

1	0.17	11	0.15
2	0.50	12	0.02
3	0.75	13	0.22
4	0.25	14	0.006
5	0.02	15	0.0004
6	0.09	16	7.0
7	1.00	17	10.00
8	2.00	18	0.14
9	1.11	19	0.115
10	0.015	20	0.2267

Chapter 4.2

1	10%	11	1%
2	20%	12	11%
3	15%	13	75%
4	30%	14	100%
5	45%	15	200%
6	62%	16	150%
7	50%	17	300%
8	74%	18	62.5%
9	9%	19	38%
10	5%	20	1%

Chapter 4.3

1	.25 = 25%	11	.1429 = 14.29
2	.75 = 75%	12	.45 = 45%
3	.20 = 20%	13	.02 = 2%
4	.40 = 40%	14	.5 = 50%
5	.60 = 60%	15	.9 = 90%
6	.10 = 10%	16	.7 = 70%
7	.30 = 30%	17	.35 = 35%
8	.125 = 12.5%	18	.15 = 15%
9	.375 = 37.5%	19	.05 = 5%
10	.625 = 62.5%	20	.333 = 33.3%

Chapter 4.4

1	3	9	3
2	15	10	300
3	2.5	11	9
4	0.3	12	24
5	12	13	9
6	25	14	1
7	20.4	15	24
8	70	16	24

17 63

18 24

19 425

20 180

Chapter 4.5

1   =

2   =

3   ≠

4   ≠

5   =

6   ≠

7   ≠

8   ≠

9   =

10 ≠

11 ≠

12 =

13 =

14 =

15 ≠

16 =

17 =

18 =

19 =

20 ≠

Chapter 4.6   TEST Chapter 4

Fraction	Ratio	%
1/2	1:2	50%
1/4	1:4	25%
3/4	3:4	75%
1/5	1:5	20%
2/5	2:5	40%
1 /1 0	1:10	10%
3/10	3:10	30%
3/5	3:5	60%
1/1	1:1	100%
2/5	4:10	40%
3	3:1	200%
3/5	6:10	60%
3/50	3:50	6%
9/20	9:20	45%
11/20	11:20	55%
20/20	20:20	100%
1/5	10:50	20%
19/50	19:50	38%
1/100	1:100	1%
1/20	5:100	5%
1/10	1:10	10%

#	Answer
1	15%
2	73%
3	21%
4	100%
5	5%
6	0.14
7	0.50
8	0.20
9	2.50
10	0.01
11	30
12	10
13	25
14	10
15	20
16	1/10
17	3/20
18	35/100 = 7/20
19	50/100 = 1/2
20	75/100 = 3/4

CHAPTER 5

GEOMETRY:

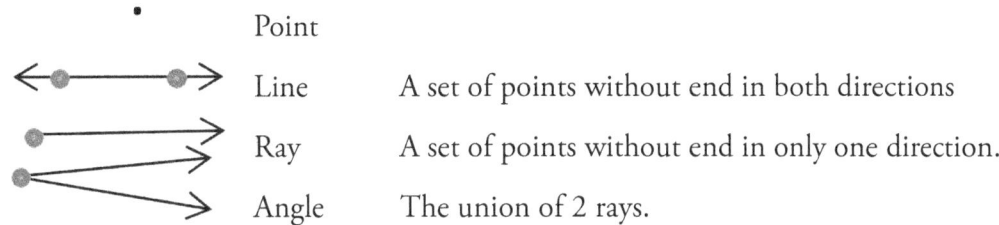

Point

Line   A set of points without end in both directions

Ray   A set of points without end in only one direction.

Angle   The union of 2 rays.

FACTS TO KNOW:

- Angles are measured with the use of a **protractor.**
- The standard unit of measurement is degrees.
- The Degree symbol is °
- An angle that measures less than 90° is called an <u>acute angle.</u>
- An angle that measures exactly 90° is called a <u>right angle.</u>
- An angle that measures more than 90° is called an <u>obtuse angle.</u>
- The 2 rays of an angle are called the <u>arms.</u>
- The point where the 2 rays meet is called the <u>vertex.</u>

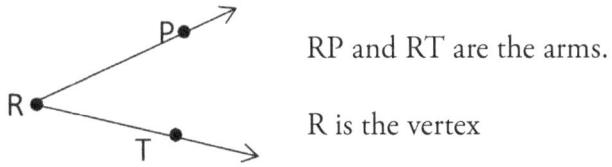

RP and RT are the arms.

R is the vertex

Use 3 letters to name an angle. The middle letter **always** refers to the vertex. The above angle would be named PRT or TRP.

PARALLEL, INTERSECTING AND PERPENDICULAR LINES:

Two lines that would never meet each other are called **Parallel Lines.**

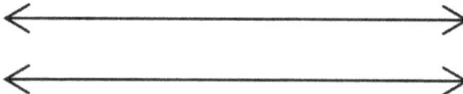

Two lines that meet each other are called **intersecting lines.**

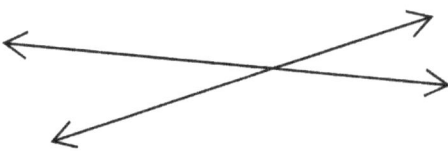

Two intersecting lines that meet each other to form right angles (90°) are called **Perpendicular Lines.**

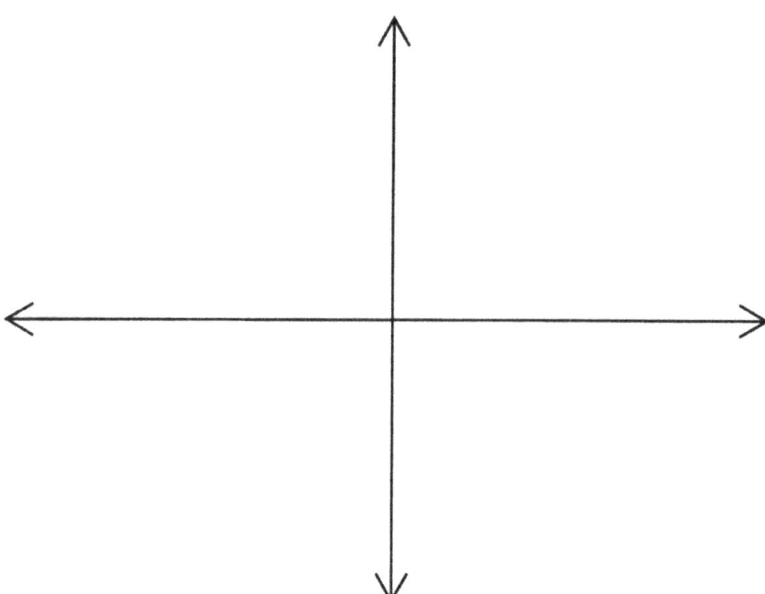

POLYGONS:

A figure that is closed with straight lines is called a polygon. "Poly" means many—"gon" means side.

The smallest number of sides a polygon can have is 3. Listed below is a partial list of the names of different polygons and the number of their sides.

3	Triangle
4	Quadrilateral
5	Pentagon
6	Hexagon
7	Heptagon
8	Octagon
9	Nonagon
10	Decagon
11	Undecagon
12	Dodecagon

TRIANGLES:

All triangles have 3 sides and 3 angles.

All triangles have a total of 180° in their 3 angles.

A triangle that has <u>all 3 sides the same length</u> is called an <u>equilateral triangle</u>.

A triangle that has <u>2 sides equal</u> is called an <u>isosceles triangle</u>.

A triangle with <u>no sides equal</u> is called a <u>scalene triangle</u>.

5.1 Problems—Name the triangles below:

1.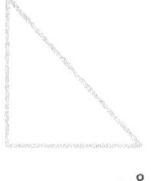

2.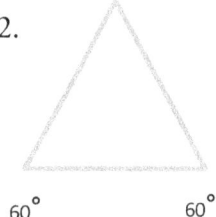

3.

4.   All triangles have 180° .   True or False?

5.   All Triangles have 4 sides.   True or False?

6.   If an equilateral triangle has one side that is 10" long, what are the lengths of the other 2 sides?

7.   If a triangle has 2 angles that total 110°, how many degrees is the third angle?

QUADRILATERALS:

A four-sided polygon is called a <u>quadrilateral</u>:

A four-side polygon is called a quadrilateral. ("Quad" Means 4.) The most common quadrilaterals are squares and rectangles. Some others are:

 Parallelogram—A quadrilateral with opposite sides that are congruent (same length) and parallel.

 Rhombus—A parallelogram with equal sides Having 2 obtuse (angle that is greater than 90° but less than 180°)angles and 2 acute (an angle that is less than 90°) angles.

 Trapezoid—A quadrilateral with only 2 parallel sides.

 Square—A quadrilateral with 4 right angles and 4 congruent sides.

 Rectangle—A quadrilateral that has 4 right angles and has opposite sides that are congruent.

PERIMETER:

To find the perimeter of any polygon, just find the distance around it. Perimeter means distance around:

In problems dealing with squares, you will usually be given only the length of one side, but since it is a square the other three sides have to be the same length.

Example: Find the perimeter of a square that has a side 10" long.

 The perimeter would be 40".

In problems dealing with rectangles you will be given one of the lengths and one of the widths. The other length and the other width is the same measurement.

Example: Find the perimeter of a rectangle that has a length of 5" and a width of 3".

 5 + 3 + 5 + 3 = 16" perimeter.

There are 2 formulas you can use for the perimeter of rectangles: (You can use either formula)

P = 2L + 2W   P = 2(L + W)

Using the example above:

P = 2(5) + 2(3)   P = 2(5 + 3)
P = 10 + 6   P = 2(8)
P = 16   P = 16

5.2 Problems—Find the perimeter of the following:

<u>Rectangles</u>   <u>Squares</u>

	L	W	P		L	W	P		S
1.	10	3		6.	7	4		11. S	6
2.	2	1		7.	8	2		12. S	10
3.	5	4		8.	100	50		13. S	11
4.	11	10		9.	40	10		14. S	½
5.	9	3		10.	10	1		15. S	1

AREA OF TRIANGLES, RECTANGLES AND SQUARES:

The area of a polygon is the amount of square units that will fit inside the figure. You could count them on most polygons but this would take too long. For triangles, rectangles and squares we have the following formulas:

Triangle	Area = base X height X	½ or	b X h X ½
Rectangles	Area = Length X Width	or	l X w
Squares	Area = Side X Side	or	s X s

To find the area of a triangle, multiply the base of the triangle times the height and then times ½.

Example: 5"   A = b x h x ½
           A = 10" x 5" x ½
  Height   A = 50" x ½
      10"   A = 25 square inches base

Your answer when working in area will always be in <u>square units.</u>
To find the area of a square, multiply one of the sides times another side.

Example: A = side x side A = 7" x 7"
           A = 49 square inches
  7"

To find the area of a rectangle, multiply one of the lengths times one of the widths.

Example: A = L x W
  6"   A = 11" x 6"
  Width   A = 66 square inches
      11"
    Length

The above answers (25, 49, 66) means that you could put that many square inches <u>inside</u> these figures.

Remember—A square inch is a square, one inch on all 4 sides.

5.3 Problems—Find the area of the following:

Triangles			Rectangles			Squares	
$A = b \times h \times \frac{1}{2}$			$A = L \times W$			$A = S^2$ or $(S \times S)$	
1.	b = 5	h = 2	11.	L = 10	W = 3	21.	S = 5
2.	b = 10	h = 4	12.	L = 9	W = 2	22.	S = 10
3.	b = 11	h = 4	13.	L = 7	W = 5	23.	S = 11
4.	b = 7	h = 2	14.	L = 2	W = 1	24.	S = 7
5.	b = 6	h = 10	15.	L = 3	W = 2	25.	S = 3
6.	b = 3	h = 2	16.	L = 11	W = 10		
7.	b = 9	h = 2	17.	L = 100	W = 50		
8.	b = 1	h = 2	18.	L = 5	W = 3		
9.	b = 2	h = 20	19.	L = 6	W = 5		
10.	b = 6	h = 2	20.	L = 15	W = 10		

CIRCLES:

A circle is a round, closed curve that has a center and every line drawn from the center is the same measurement. All circles have 360 degrees (360°). A bicycle wheel is an example of a circle. The spokes on the wheel are examples of radii (singular of radius).

FACTS TO LEARN:

Radius—A line segment extending from the center of a circle to any point on the circle. "OA" below is an example.

Diameter—A line segment extending from one point on a circle to another point on the circle and passing through the center. "AOB" below is an example.

Cord—A line segment that touches at two points within the circle. "XY" below is an example.

Compass—A Mathematical instrument used to draw circles.

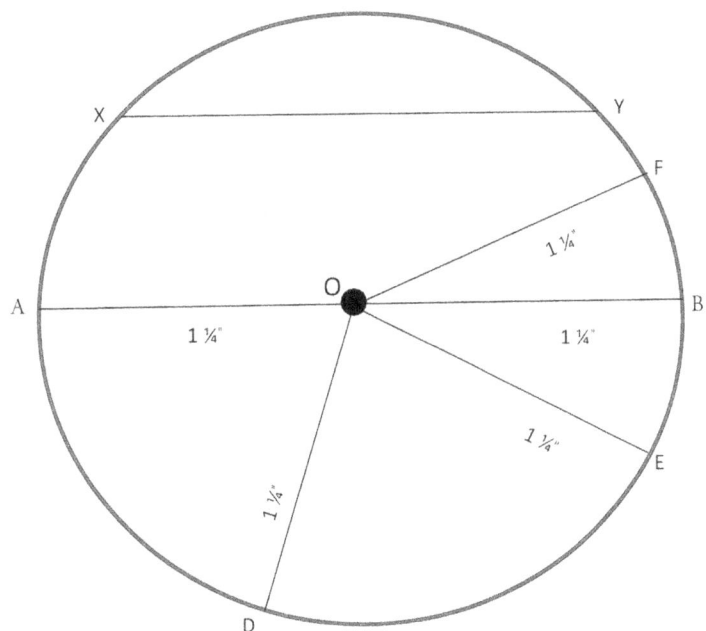

CIRCUMFERENCE OF A CIRCLE:

When we find the distance around a polygon, we call it perimeter. When we find the distance around a circle we call it the circumference. Since it is very difficult to measure curved lines we must use a formula to find the circumference of a circle. The formula is C = π d, which means that the circumference (C) is equal to pi (π), times the diameter (d). Pi is always 3.14 and the diameter will be given.

Pi (π) is a Greek letter that stands for the relationship of the circumference to the diameter. Mathematicians discovered many years ago that the distance around a circle is a little more than 3 times the distance across it. If we know the distance across a circle we can find the distance around it by multiplying the diameter times Pi (π).

Example: Find the circumference of a circle that is 11" in diameter.

$$C = \pi \times d \qquad\qquad\qquad 3.14$$
$$C = 3.14 \times 11" \qquad\qquad\quad \underline{\times 11}$$
$$C = 34.54 \qquad\qquad\qquad\quad 3.14$$
$$\underline{31.4}$$
$$34.54$$

A circle that is 11" across is 34.54" around. If we were given a problem in circumference, you will need to know the diameter. Sometimes we are not given the diameter but are given the radius. If this happens, simply double the radius (2 radii = 1 diameter) and multiply by Pi (π).

Example: Find the circumference of a circle when the radius is 6" If R = 6", then D = 12"

$$C = \pi \times D \qquad\qquad\qquad 3.14$$
$$C = 3.14 \times 12" \qquad\qquad\quad \underline{12}$$
$$C = 37.68" \qquad\qquad\qquad\quad 6.28$$
$$\underline{31.4}$$
$$37.68$$

5.4 Problems—Find the circumference of a circle:

1. D = 4   9. D = 7

2. D = 10   10. D = 11

3. D = 20   11. R = 8

4. D = 6   12. R = 20

5. D = 3   13. R = 6

6. D = 2   14. R = 100

7. D = 1   15. R = 24

8. D = 5

AREA OF A CIRCLE:

To find the number of squares that will fit inside a circle it is impractical to actually count them. We have to use the formula, A =   R², which means the area of a circle is equal to pi () times R² (R x R).

Example—Find the area of a circle that has a radius of 5":

Area = π x R x R	3.14
A = 3.14 x 5 x 5	x 25
A = 3.14 x 25	15.70
A = 78.50 square Inches	62.8
	78.50

A circle that has a radius of 5" will have an area of 78.50 square inches.

If the diameter is given instead of the radius, take ½ of the diameter to find the radius, then use the formula.

Example—Find the area of a circle that has a diameter of 14":

If D = 14", then R =7" A = π x R x R

A = 3.14 x 7 x 7
A = 3.14 x 49
A = 153.86 square inches

5.5 Problems—Find the area of these circles:

1.   R = 6   6.   R = 10

2.   R = 3   7.   R = 7

3.   R = 2   8.   R = 11

4.   R = 1   9.   D = 10

5.   R = 5   10.   D = 20

TEST—CHAPTER V—GEOMETRY

Find the area of the circles below: $A = \pi R^2$

    1.   R = 6"
    2.   R = 3"
    3.   D = 10"

Find the circumference of the circles below: $C = \pi D$:

    4.   D = 4"
    5.   D = 10"
    6.   R = 16"

Find the perimeter of the following: Rectangles:

    7.   L = 10"   W = 3"
    8.   L = 9"   W = 2"

Squares:

    9.   S = 10"
    10.   S = 6"

Find the area of the following: Triangles

    11.   B = 7"   H = 2"
    12.   B = 11" H = 4"

Rectangles

    13.   L = 9"   W = 2"
    14.   L = 3"   W = 2"
    15.   S = 10"

True or False:

    16.   _____The union of 2 rays is called an angle.

    17.   _____Angles are measured with a compass.

    18.   _____The standard unit of measurement for angles is minutes.

    19.   _____A set of points without end in both directions is called a line.

    20.   _____An angle that measures exactly 90° is called an acute angle.

21. _____An angle that measures more than 90° is called an obtuse angle.

22. _____An 8-sided polygon is called an octagon.

23. _____Two lines that would never meet each other are called parallel lines.

24. _____All circles have 360° .

25. _____All triangles have 190° .

26. _____A triangle that has all 3 sides the same length is called an scalene triangle.

27. _____A parallelogram with equal sides, having 2 obtuse angles and 2 acute angles, is called a rectangle.

28. _____All Radii of a given circle are equal.

29. _____Pi (π) is the distance across a circle.

30. _____All Polygons are 4 sided.

31. _____All squares are quadrilaterals.

32. _____All quadrilaterals are squares.

33. _____A scalene triangle has all sides equal.

ANSWERS CHAPTER 5

Chapter 5.1
1	Right	5	FALSE
2	Equilateral	6	10" each
3	Isocelese	7	70°
4	TRUE		

Chapter 5.2
1	26	9	100
2	6	10	22
3	18	11	24
4	42	12	100
5	24	13	44
6	22	14	2
7	20	15	4
8	300		

Chapter 5.3
1	5	14	2
2	20	15	6
3	22	16	110
4	7	17	5000
5	30	18	15
6	3	19	30
7	9	20	150
8	1.5	21	25
9	20	22	100
10	6	23	121
11	30	24	49
12	18	25	9
13	35		

Chapter 5.4
1	12.56	8	15.70
2	31.40	9	21.98
3	62.80	10	34.54
4	188.40	11	50.24
5	9.42	12	125.6
6	6.28	13	100.48
7	3.14	14	628
		15	150.72

Chapter 5.5

1	113.04 sq	6	314 sq
2	28.26 sq	7	153.86 sq
3	12.56 sq	8	379.94 sq
4	3.14 sq	9	78.5 sq
5	78.5 sq	10	1256 sq

TEST Chapter 5

1	113.4 sq"	18	F
2	28.26 sq"	19	T
3	78.5 sq"	20	F
4	12.56"	21	T
5	31.4"	22	T
6	100.48"	23	T
7	26"	24	T
8	22"	25	F
9	40	26	F
10	24	27	F
11	7 sq"	28	T
12	22 sq"	29	F
13	18 sq"	30	T
14	6 sq"	31	T
15	100 sq"	32	F
16	T	33	F
17	F		

CHAPTER 6

MEAN, MODE AND MEDIAN:

Mean—The mean of a group of numbers is the "arithmetic average". To calculate the mean, add the numbers and then divide the total by the number of numbers in the group.

$$
\begin{array}{r}
65 \\
70 \\
100 \\
50 \\
40 \\
\underline{81} \\
306
\end{array}
\qquad
\begin{array}{r}
51 \text{ average or mean} \\
6)\overline{306} \\
\underline{300} \\
06 \\
\underline{06} \\
0
\end{array}
$$

Mode—The mode of a group of numbers is the number represented the <u>most frequently</u>.

$$
\begin{array}{r}
65 \\
70 \\
80 \\
65 \\
\underline{100} \\
380
\end{array}
\qquad
\text{Mode} = 65
$$

Median—The median score of a group of numbers is the score that is in the <u>middle</u>. If the scores are odd in number, the median can be found by arranging the scores in order. 65, 70, 90, 60, 100:

The median is: 60

65
(70)
90
100

If the scores are even in number, arrange them in order, then find the median by adding the two middle scores together and dividing by 2. The median may not actually be in the set of numbers.

> Mean = average
> Mode = most
> Median = middle

6.1 Problems—Find the mean, mode and median:

1.	2.	3.
5	50	10
6	60	10
7	70	10
8	80	15
8	90	20
10	100	25
11	100	30

4.	5.	6.
30	70	65
32	70	65
32	80	70
32	100	70
36		70
38		80
		80
		85
		85
		100

7. 41

    42

    43

    44

    45

    <u>45</u>

8. 19

    19

    19

    20

    30

    40

    60

    100

    <u>102</u>

9. 44

    63

    72

    90

    90

    90

    <u>100</u>

10. 50

    100

    200

    300

    500

    <u>500</u>

EQUATIONS:

Sometimes it will be necessary to add or subtract terms of the equations before you can solve for the unknown letter. There are two words you should learn—variables and constants. Variables are the letters, and get there name because their value will vary in different problems. Constants are the numbers, and they get their name because their value is always the same.

$R + 6 = 12$ 6 and 12 are constants; R is the variable. Collect all variables of the same letter together and collect constants together.

$$5B + 6B = 10 + 23 \qquad\qquad 3A + A = 16$$
$$11B = 33 \qquad\qquad\qquad 4A = 16$$
$$B = 3 \qquad\qquad\qquad\quad A = 4$$

6.2 Problems—Collect all terms and solve for the unknown:

1. $3A = 10 + 5$

2. $4Y + 2Y = 12$

3. $5C + C = 6$

4. $30 = 5B + 10B$

5. $15 + 10 = 5C$

6. $4x = 10 + 2$

7. $3X + 4X = 7 + 21$

8. $14 + 6 = 9X + X$

9. $6Q = 10 + 20$

10. $7Y + Y = 16$

11. $4C + 4C = 80$

12. $10C = 10 + 20$

13. $7B = 1 + 6$

14. $3W = 15 + 6$

15. $B + B = 6$

EQUATIONS—SOLVING FOR UNKNOWNS:

If 5 pencils cost 20 cents, then 1 pencil costs 4 cents. This is a problem where you know how much 5 of something cost, and you figured out how much one cost by dividing 20 cents by 5 pencils = 4 cents. This is similar to solving equations.

Example: If 6 apples = 12 cents, how much is one apple?   2 cents.

$6A = 12$ what is the value of A?

THINK: What number can you put in place of "A" and when you multiply it by 6 you will get 12? The answer is 2.

$6 (2) = 12$

$6 \times 2 = 12$ When two numbers are written together and one is in parentheses, it is understood that they are to be multiplied together.

Another way to work this problem is to divide. We know that if 6 A's = 12, to find out what one A is, we can divide both sides by 6.

$$6A = 12$$
$$\frac{6A}{6} = \frac{12}{6}$$
$$\frac{1A}{1} = \frac{2}{1}$$
$$A = 2$$

6.3 Problems—Solve for the unknown letter.

1. 6X = 12

2. 5A = 5

3. 3Y = 9

4. 7X = 21

5. 4Q = 20

6. 2R = 6

7.   2X = 10

8.   3Q = 12

9.   7Y = 14

10. 10Y = 100

11. 10Y = 20

12. 7R = 7

13. 4X = 20

14. 11X = 121

15. 10Y = 200

16. 4X = 40

17. 5C = 55

18. 6B = 42

19. 7R = 49

20. 10X = 30

METRIC SYSTEM:

10^4	10^3	10^2	10^1	10^0	10^{-1}	10^{-2}	10^{-3}	10^{-4}
M E G A M E T E R	K I L O M E T E R	H E C T O M E T E R	D E C A M E T E R	M E T E R	D E C I M E T E R	C E N T I M E T E R	M I L L I M E T E R	M I C R O M E T E R

To change from any unit in the metric system, simply move the decimal point to the right or the left the number of "steps" it takes to go from one unit to the other.

    3 meters = 30 decimeters
    3 meters = .3 decameters
    6.5 hectometers = 650 meters

6.4 Problems—Change the following:

1. 4 meters to decimeters_____

2. .6 meters to decameters_____

3. .7 hectometers to meters_____

4. 7 decimeters to meters_____

5. 7 kilometers to meters_____

6. 11 meters to kilometers_____

7. 1.7 meters to kilometers_____

8. 22 meters to decimeters_____

9. 5 centimeters to meters_____

10. 7 meters to decimeters_____

11. 14 micrometers to meters_____

12. 9 Kilometers to meters_____

13. 3 meters to centimeters_____

13. 3 meters to centimeters_____

14. 4 decimeters to meters_____

15. 6 dekameters to decimeters_____

To change from the Metric System to the British—American System, use the table below:

When you know	Multiply By	To Find
Inches	25.4	Millimeters
Yards	.91	Meters
Miles	1.609	Kilometers
Square Yards	.836	Square Meters
Acres	.405	Hectares
Cubic Yards	.765	Cubic Meters
Quarts	.946	Liters
Ounces	28.350	Grams
Pounds (AVDP)	.454	Kilograms
Fahrenheit	5/9 (after sub. 32)	Celsius Temperature
Millimeters	.039	Inches
Meters	3.281	Feet
Meters	1.094	Yards
Kilometers	.621	Miles
Square Meters	1.196	Square Yards
Hectares	2.471	Acres
Cubic Meters	1.308	Cubic Yards
Liters	1.057	Quarts
Grams	.035	Ounces
Kilograms	2.205	Pounds
Celsius Temperature	9/5 (and add 32)	Fahrenheit Temperature

6.5 Problems—Convert the following measurements:

1. 2 inches to _____ Millimeters.

2. 10 Millimeters to _____ Inches.

3. 100 Yards to _____ Meters.

4. 5 Miles to _____ Kilometers.

5. 100 Meters to _____ Yards.

6. 5 Kilometers to _____ Miles.

7. 121 Millimeters to _____ Inches.

8. 80 Fahrenheit to _____ Celsius.

9. 50 Celsius to _____ Fahrenheit.

10. 16 Ounces to _____ Grams.

11. 20 Grams to _____ Ounces.

12. 4 Quarts to _____ Liters.

13. 120 Meters to _____ Yards.

14. 13 Pounds to _____ Kilograms.

15. 3 Feet to _____ Meters.

NON DECIMAL BASES:

In this section we will briefly study 3 bases, other than base 10. Our system (decimal numbers) is based on powers of 10's. Every place-value increases by a multiple of 10. The number 13,964 means: $\underline{1}$ ten thousand, $\underline{3}$ thousands, $\underline{9}$ hundreds, $\underline{6}$ tens, and $\underline{4}$ ones. We will learn base 2, 5 and 7. There are other bases but these will be learned later in school.

Base 2—Using base 2 as a system of counting, every place-value increases by powers of 2. Below is a chart showing the number 125 in base 10 and also in base 2. [2]

125 =

1	2	5
10^2	10^1	10^0

125 =

1	1	1	1	1	0	1
2^6	2^5	2^4	2^3	2^2	2^1	2^0
64	32	16	8	4	2	1

1	64
1	32
1	16
1	8
1	4
0	2
1	1
	125

In base 10 we have 10 different numerals to use: 0–9. In base 2 we only have 2 numerals to use: 0 and 1. Remember: Any number raised to the zero power is equal to 1.

6.6 Problems: Change to base 2.

1.   17

2.   100

3.   65

4.   9

5.   7

NON DECIMAL BASES: (CONTINUED)

6.6 Problems: Change to base 10.

6. 101_2 9. 1000_2

7. 11_2 10. 1001_2

8. 1111_2

Base 5 and Base 7.

Base 5 increases by multiples of 5.

5^4	5^3	5^2	5^1	5^0
625	125	25	5	1

Base 7 increases by multiples of 7.

7^4	7^3	7^2	7^1	7^0
2401	343	49	7	1

Always start the table with the base in zero power. (2^0, 10^0, 5^0, 7^0)

You can now see the other bases (3, 4, 6, 8, 9, 11, 12) would be worked in the same manner.
In base 5 we have 5 numerals (0, 1, 2, 3, 4).
In base 7 we have 7 numerals (0, 1, 2, 3, 4, 5, 6).
In base 2 we have 2 numerals (0,1).
In any base the highest numeral is always one less than the base.

6.7 Problems—Change to the various bases:

1. 101_2 to base 5

2. 23_5 to base 10

3. 121_7 to base 10

4.   23 to base 7

5.   145 to base 5

6. 1101_2 to base 10

7. 33_7 to base 5

8. 33_5 to base 7

9. 140_5 to base 10

10. 13_7 to base 10

11. 101_2 to base 10

12. 121_5 to base 2

TEST—CHAPTER VI—MEAN, MODE & MEDIAN, EQUATIONS, METRIC AND BASES

Solve for the unknown:

1. $3Y = 9$
2. $2X = 10$
3. $4X = 20$
4. $7R = 49$
5. $10X = 30$
6. $4Y + 2Y = 12$
7. $30 = 5B + 10B$
8. $6Q = 10 + 20$

Change to base 2

9. 23
10. 171
11. 121_5

Change to base 10

12. 1101_2

13. 34_5

Find the mean, median and mode:

14. mean=
15. median=
16. mode=

30
35
45
75
90
90
100

Change the following:

17. 3 meters to _____ decimeters.

18. 11 kilometers to _____ meters.

19. .6 meters to _____ decameters.

20. 4 decimeters to _____ hectometers.

ANSWERS—CHAPTER VI

Chapter 6.1

	Mean	Mode	Median
1	7.86	8	8
2	78.60	100	80
3	17.14	10	15
4	33.30	32	32
5	80.00	70	75
6	77.00	70	75
7	43.33	45	43.5
8	45.40	19	30
9	78.43	90	90
10	27.50	500	250

Chapter 6.2

1. a=5
2. y=2
3. c=1
4. b=2
5. c=5
6. x=3
7. x=4
8. x=2
9. q=5
10. y=2
11. c=10
12. c=3
13. b=1
14. w=7
15. b=3

Chapter 6.3

1. x=2
2. a=1
3. y=3
4. x=3
5. q=5
6. r=3
7. x=5
8. q=4
9. y=2
10. y=10
11. y=2
12. r=1
13. x=5
14. x=11
15. y=2
16. x=10
17. c=11
18. b=7
19. r=7
20. x=3

Chapter 6.4

1	40	9	.05
2	.06	10	70
3	70	11	.0014
4	.7	12	9000
5	7000	13	300
6	.011	14	.04
7	.0017	15	600
8	220		

Chapter 6.5

1	508	9	122.000
2	0.390	10	453.600
3	91.000	11	0.700
4	8.045	12	3.784
5	109.400	13	131.280
6	3.105	14	5.902
7	4.719	15	0.910
8	26.400		

Chapter 6.6

1	10001_2	6	5
2	100000_2	7	3
3	1000001_2	8	15
4	1001_2	9	8
5	101_2	10	9

Chapter 6.7

1	1_5	7	41_5
2	13	8	24_7
3	50	9	45
4	32_7	10	10
5	1040_5	11	5
6	23	12	100100_2

TEST Chapter 6

1	y=3	11	55_5
2	x=5	12	13
3	x=5	13	95
4	r=7	14	66.43
5	x=3	15	75
6	y=2	16	90
7	b=2	17	30
8	q=5	18	11000
9	10111_2	19	0.06
10	10101011_2	20	0.4

Chapter 1.1

1	11		11	40
2	8		12	29
3	4		13	95
4	13		14	42
5	17		15	52
6	23		16	60
7	27		17	75
8	30		18	87
9	24		19	1901
10	44		20	1776

Chapter 1.2

1	IX		14	XCIX
2	XI		15	LI
3	XIV		16	CD
4	XVI		17	DC
5	XIX		18	MMXV (2015)
6	XXI		19	YOUR HOUSE #
7	XXI		20	YOUR DoB
8	XXIV		21	CM
9	XVI		22	MC
10	XXXIX		23	DXXXV
11	XLI		24	<u>CL</u>MMM
12	LI		25	<u>LX</u>MMDXLV
13	XLLVI			

Chapter 1.3

1	Example
2	2 Hundreds, 3 Tens, 6 Ones
3	7 Hundreds, 9 Tens, 1 Ones
4	8 Thousands, 0 Hundreds, 5 Tens, 9 Ones
5	1 Thousand, 1 Hundred, 0 Tens, 1 Ones
6	2 Thousands, 0 Hundreds,) Tens, 7 Ones
7	5 Hundreds, 6 Tens, 2 Ones
8	4 Hundreds, 9 Tens, 0 Ones
9	3 Thousands, 1 Hundred, 2 Tens, 3 Ones
10	7 Thousands, 1 Hundred, 9 Tens, 9 Ones
11	3 Tens, 7 Ones

12   9 Tens, 9 Ones

13   9 Hundreds, 9 Tens, 9 Ones

14   2 Ten Thousands, 9 Thousands, 0 Hundreds, 9 Tens, 9 Ones

15   1 Ten Thousands, 0 Thousands, 7 Hundredths, 6 Tens, 1

16   1 Hundred Thousand, 2 Ten Thousands, 3 Thousands, 6 Hundreds, 7 Tens, 9 Ones

Chapter 1.4

1	Even	6	Odd
2	Odd	7	Odd
3	Even	8	Odd
4	Even	9	Even
5	Even	10	Odd

Chapter 1.5

1	56,565	11	98,654
2	797,113	12	1,001
3	8,900	13	77,787,771
4	10,000	14	234,671
5	10,000,000	15	88,731,109
6	100,000,000	16	43,190,678,914
7	711,900,718	17	942,286,002
8	29,801,236	18	11,171
9	1,239,347	19	7,112,461
10	320,761	20	6,667,780,000

Chapter 1.6

1)
```
      6
     60
 +  600
    666
```

2)
```
      1
    203
   3711
 +   97
   4012
```

3)
```
     14
    148
 +   73
    235
```

4)
```
    560
     17
 +   64
    641
```

5)
```
     99
 +    1
    100
    641
```

6)
```
     76
    112
 +    7
    195
```

7)
```
   4033
    999
 +   70
   5102
```

8)
```
      1
     11
    111
 + 1111
   1234
```

9)
```
    365
   5630
 +    2
   5997
```

10)
```
    114
   4191
      1
 +   97
   4403
```

11)   367
      − 24
      343

12)   999
      − 10
      989

13) 1000
     − 999
        1

14)   75
      − 5
      70

15)   83
      − 38
      45

16) 5007
     − 93
    4914

17) 4100
     − 909
    3191

18)   781
      − 92
     689

19) 5321
     − 971
    4350

20)   423
      − 62
     361

Chapter 1.7

1 $7 = 6 + 1$
2 $8 \times 1 = 8$
3 $9 + 0 = 9$
4 $4 - 2 < 2 + 1$
5 $16 + 1 = 17$
6 $9 \times 2 = 20 - 2$
7 $1 + 10 > 11 \times 0$
8 $6 - 1 < 10 - 4$
9 $15 + 15 > 50 -$
10 30
11 $100 - 9 = 90 + 1$

12 $31 + 2 < 42 \times 1$
13 $17 - 9 > 7$
14 $100 + 1 > 100 \times$
15 1
16 $99 - 9 = 90$
17 $44 \times 1 = 50 - 6$
18 $73 - 4 = 69 \times 1$
19 $7 \times 1 > 17 \times 0$
20 $49 + 49 = 98 \times 1$
21 $23 + 32 = 60 - 5$
22 $14 + 1 > 15 - 1$

Chapter 1.8

1	11 < 12		6	14 < 13 + 3
2	12 > 11		7	100 < 90 x 2
3	6 + 1 < 8		8	5 + 5 + 4 > 10
4	9 - 1 > 7		9	19 > 20 - 2
5	2 x 3 > 5		10	16 x 1 < 17

Chapter 1.9

1	78		7	99
2	130		8	171,579
3	915		9	4,784
4	1,430		10	763
5	14,248		11	8,390
6	1,911		12	2,100

Chapter 1.10

1	8		14	139
2	1		15	460
3	32		16	247
4	23		17	460
5	98		18	1,000
6	40		19	700
7	10		20	3,319
8	36		21	112
9	1		22	111
10	111		23	320
11	131		24	20
12	464		25	889
13	511			

Chapter 1.11

1	8		11	990
2	45		12	396
3	63		13	1088
4	2		14	999
5	5		15	1
6	9		16	25,903
7	210		17	163
8	143		18	900
9	202		19	99
10	49		20	110

Chapter 1.12

1	90	16	196
2	66	17	4,527
3	120	18	6,118
4	454	19	240
5	126	20	4,080
6	51	21	11,628
7	48	22	45,881
8	156	23	149,314
9	1,848	24	9,450
10	4,536	25	10,989
11	924	26	12,321
12	420	27	1,250
13	930	28	2,346
14	84	29	6,060
15	774	30	44,880

Chapter 1.13

1	3	11	62
2	8	12	21
3	10	13	91
4	13	14	23
5	3	15	26 R2
6	7	16	205
7	6	17	6 R1
8	8	18	10 R1
9	9	19	14 R2
10	8	20	5 R2

Chapter 1.14

1	0.07	15	0.0043
2	6.2	16	0.017
3	0.112	17	42.5
4	1.7	18	41
5	400	19	0.71
6	4.2	20	0.037
7	11.1	21	1.121
8	73.1	22	2.2
9	6,110	23	0.22
10	0.087	24	0.0022
11	7.7	25	0.6
12	0.77	26	73
13	6.3	27	12.1
14	1.9		

Chapter 1.15

1	2,4,8	11	2,4,8
2	3,5	12	3,9
3	5	13	3
4	2,4,8,10	14	3
5	2,4,5,10	15	-
6	3,7	16	2,4,8
7	2,4,8	17	2
8	3,5	18	2,3,4,6,9
9	2,3,4,6	19	2
10	2,4	20	2,3,4,5,6,9

Chapter 1.16

1	16	9	13
2	17	10	175
3	17	11	0
4	0	12	U
5	0	13	19
6	4	14	19
7	125	15	21
8	13		

Chapter 1.17

Numbers	Power	Base	Value
2^3	3	2	8
6^1	1	6	6
7^0	0	7	1
16^2	2	16	256
3^2	2	3	9
4^2	2	4	16
5^4	4	5	625
6^2	2	6	36
8^1	1	8	8
10^1	1	10	10
12^2	2	12	144
11^2	2	11	121
173^0	0	173	1
6^3	3	6	216
3^5	5	3	405
1^9	9	1	1
15^1	1	15	15
192^0	0	192	1
2^7	7	2	128
3^3	3	3	27

Chapter 1
TEST

1	210		14	3,737
2	443		15	3,960
3	148		16	8
4	1,428		17	5
5	1,397		18	11
6	8		19	12R2
7	50		20	12
8	12		21	9
9	389		22	45
10	467		23	XC
11	99		24	XL
12	51		25	3,207
13	5,922			

Chapter 2.1

1	2
2	9

Chapter 2.2

1

1	21	36	55	74	90
4	22	38	56	75	92
6	24	39	58	76	94
8	25	40	60	77	95
9	26	42	62	78	96
10	27	44	64	80	98
12	28	45	65	81	99
14	30	26	66	82	100
15	32	47	68	84	
16	33	50	69	85	
18	34	52	70	86	
20	35	54	72	88	

Chapter 2.3

1	=		11	=
2	=		12	≠
3	≠		13	≠
4	≠		14	≠
5	≠		15	=
6	≠		16	≠
7	=		17	=
8	=		18	≠
9	=		19	≠
10	=		20	≠

Chapter 2.4

1	2/3		8	1/5
2	1/2		9	1/2
3	2/3		10	2/3
4	1/3		11	3/4
5	1/2		12	3/5
6	5/6		13	1/4
7	1/5		14	1/3

15	3/4		18	1/5
16	1/3		19	1/10
17	2/3		20	7/10

Chapter 2.5

1	6		6	5 2/9
2	5		7	3
3	6 2/5		8	2 1/2
4	3 1/2		9	1 1/2
5	2 1/4		10	3 1/2

Chapter 2.6

1	3/2		11	24/7
2	5/2		12	27/5
3	19/74		13	14/3
4	91/9		14	7/2
5	7/2		15	5/2
6	17/3		16	5/3
7	52/7		17	39/5
8	21/2		18	67/10
9	34/3		19	11/6
10	14/6		20	11/4

Chapter 2.7

1	12		11	6
2	10		12	6
3	8		13	6
4	4		14	3
5	10		15	5
6	12		16	14
7	12		17	12
8	14		18	20
9	90		19	5
10	10		20	5

Chapter 2.8

1	1		11	9/11
2	2/3		12	1
3	2/5		13	1
4	4/5		14	11/15
5	1/2		15	1
6	6/7		16	5/6
7	4/5		17	2/5
8	7/12		18	3/7
9	1/2		19	12/19
10	7/9		20	22/25

Chapter 2.9

1	4/8 + 2/8 = 6/8 = 3/4	11	5/16
2	7/12	12	2/3
3	1/15 + 6/15 = 7/15	13	11/16
4	19/20	14	1/2
5	3/6 + 4/6 = 7/6 = 1 1/16	15	1 4/15
6	1 5/12	16	1 1/6
7	7/42 + 12/42 = 19/42	17	9 1/4
8	11/12	18	4 7/10
9	1/2	19	10 5/9
10	11/20	20	7 1/3

Chapter 2.10

1	1/2	11	5/9
2	2	12	1/12
3	1/3	13	3/13
4	1/2	14	5/17
5	3/7	15	1/5
6	1/5	16	1/50
7	5/8	17	2/15
8	5/11	18	1/3
9	1/5	19	7/19
10	3/5	20	11/23

Chapter 2.11

1	1/6	11	1/3
2	11/20	12	1/4
3	1/2	13	1/6
4	1/2	14	11/16
5	1/3	15	7/30
6	4/21	16	9/35
7	1/2	17	11/20
8	3/4	18	3/4
9	2/15	19	39/70
10	3/5	20	1/72

Chapter 2.12

1	6 1/3	11	4 2/25
2	4 3/5	12	4 1/50
3	5 1/2	13	2 5/17
4	4	14	2 1/3
5	1 1/3	15	1 1/5
6	7 3/7	16	5 1/3
7	10 1/5	17	3 2/7
8	11 2/5	18	6 3/8
9	1 1/20	19	2 1/6
10	1 7/11	20	7 1/10

Chapter 2.13

1	1/4	11	1/6
2	1/12	12	1/10
3	1/21	13	1/4
4	1/3	14	6/49
5	1/12	15	4/57
6	4/9	16	8/27
7	5/9	17	3/16
8	9/20	18	9/14
9	1/33	19	3/4
10	2/15	20	2/7

Chapter 2.14

1	7/18	11	10
2	1/4	12	100
3	3/20	13	12
4	7/80	14	28
5	9/50	15	15
6	1/12	16	2
7	1/21	17	1/2
8	1	18	2 2/3
9	1	19	1 1/11
10	14	20	2 1/3

Chapter 2.15

1	4 1/2	6	49 1/7
2	3 3/4	7	2
3	2 1/5	8	9 1/11
4	12 2/5	9	1 1/6
5	10 1/2	10	1 1/5

Chapter 2.16

1	9	9	200
2	24	10	11
3	200	11	12
4	105	12	3
5	52	13	3
6	21	14	5
7	41	15	5
8	37	16	10 1/5

TEST

1	4/5		14	1 5/8
2	2/3		15	28 5/12
3	6/7		16	2/3
4	11/15		17	30
5	1 1/4		18	2/21
6	1/6		19	1/3
7	2/5		20	10
8	4/9		21	2/3
9	3/4		22	1/2
10	1/6		23	3/5
11	1/4		24	9
12	21/32		25	5 2/3
13	3/4			

Chapter 3.1

   1   Six Tenths

   2   Three and One Tenth

   3   Forty-Two and Three Hundreths

   4   Sixty-Seven Hundreths

   5   One and Twelve Hundreths

   6   One and Nine Hundred Ninety-Two Thousandths

   7   Seven and Two Hundreths

   8   Fifthy-Three and Fourty-Two Hundreths

   9   Nine and Nine Hundred One Thousandths

  10   Ten and One Hundred One Thousandths

Chapter 3.2

1	0.7	6	9.09
2	3.4	7	4.006
3	61.3	8	0.52
4	17.02	9	8.6
5	0.32	10	93.07

Chapter 3.3

1	7.26	11	8.84
2	10.82	12	11.042
3	3.366	13	93.211
4	8.37	14	113.211
5	97.6	15	86.602
6	2.156	16	194.01
7	99.99	17	9.954
8	134.22	18	15.51
9	2.23	19	79.825
10	695	20	151.69

Chapter 3.4

1	0.2	11	6
2	0.2	12	13.919
3	2.45	13	0.01
4	4.1	14	82.87
5	1.02	15	7.05
6	62.1	16	5.9
7	1	17	8.36
8	0.10	18	10
9	1.39	19	0.06
10	12.2	20	42.988

Chapter 3.5

1	3.72	11	18.6
2	0.372	12	58.2
3	0.0372	13	123.06
4	0.602	14	3
5	6.02	15	0.5
6	0.014	16	0.322
7	0.0009	17	40.8
8	2	18	2.5
9	6.39	19	89.1
10	0.0108	20	0.4

Chapter 3.6

1	0.07	11	1.1
2	0.125	12	0.5
3	0.2	13	0.007
4	3.3	14	0.22
5	0.31	15	1.1
6	0.1	16	2.1
7	1.1	17	0.31
8	1.5	18	0.32
9	2.1	19	2
10	0.24	20	1.1

Chapter 3.7

1	30	11	40
2	0.3	12	0.4
3	0.3	13	5
4	30	14	60
5	300	15	0.06
6	0.3	16	30
7	210	17	3.1
8	21	18	2.1
9	0.8	19	1000
10	0.08	20	105

TEST Chapter 3

1	4.4	11	0.0143
2	11.21	12	0.0081
3	148.81	13	69
4	11.143	14	0.0244
5	1.144	15	389.4
6	4.63	16	20
7	7.7	17	0.2
8	6.926	18	0.2
9	0.21	19	20
10	0.222	20	2

Chapter 4.1

1	0.17	11	0.15
2	0.50	12	0.02
3	0.75	13	0.22
4	0.25	14	0.006
5	0.02	15	0.0004
6	0.09	16	7.0
7	1.00	17	10.00
8	2.00	18	0.14
9	1.11	19	0.115
10	0.015	20	0.2267

Chapter 4.2

1	10%	11	1%
2	20%	12	11%
3	15%	13	75%
4	30%	14	100%
5	45%	15	200%
6	62%	16	150%
7	50%	17	300%
8	74%	18	62.5%
9	9%	19	38%
10	5%	20	1%

Chapter 4.3

1	.25 = 25%	11	.1429 = 14.29
2	.75 = 75%	12	.45 = 45%
3	.20 = 20%	13	.02 = 2%
4	.40 = 40%	14	.5 = 50%
5	.60 = 60%	15	.9 = 90%
6	.10 = 10%	16	.7 = 70%
7	.30 = 30%	17	.35 = 35%
8	.125 = 12.5%	18	.15 = 15%
9	.375 = 37.5%	19	.05 = 5%
10	.625 = 62.5%	20	.333 = 33.3%

Chapter 4.4

1	3	7	20.4
2	15	8	70
3	2.5	9	3
4	0.3	10	300
5	12	11	9
6	25	12	24

13	9		17	63
14	1		18	24
15	24		19	425
16	24		20	180

Chapter 4.5

1	=		11	≠
2	=		12	=
3	≠		13	=
4	≠		14	=
5	=		15	≠
6	≠		16	=
7	≠		17	=
8	≠		18	=
9	=		19	=
10	≠		20	≠

Chapter 4.6   TEST Chapter 4

Fraction	Ratio	%
1/2	1:2	50%
1/4	1:4	25%
3/4	3:4	75%
1/5	1:5	20%
2/5	2:5	40%
1 /1 0	1:10	10%
3/10	3:10	30%
3/5	3:5	60%
1/1	1:1	100%
2/5	4:10	40%
3	3:1	200%
3/5	6:10	60%
3/50	3:50	6%
9/20	9:20	45%
11/20	11:20	55%
20/20	20:20	100%
1/5	10:50	20%
19/50	19:50	38%
1/100	1:100	1%
1/20	5:100	5%
1/10	1:10	10%

1   15%
2   73%
3   21%
4   100%
5   5%
6   0.14
7   0.50
8   0.20
9   2.50
10   0.01
11   30
12   10
13   25
14   10
15   20
16   1/10
17   3/20
18   35/100 = 7/20
19   50/100 = 1/2
20   75/100 = 3/4

Chapter 5.1
1 Right
2 Equilateral
3 Isocelese
4 TRUE
5 FALSE
6 10" each
7 70°

Chapter 5.2
1 26
2 6
3 18
4 42
5 24
6 22
7 20
8 300
9 100
10 22
11 24
12 100
13 44
14 2
15 4

Chapter 5.3
1 5
2 20
3 22
4 7
5 30
6 3
7 9
8 1.5
9 20
10 6
11 30
12 18
13 35
14 2
15 6
16 110
17 5000
18 15
19 30
20 150
21 25
22 100
23 121
24 49
25 9

Chapter 5.4
1 12.56
2 31.40
3 62.80
4 188.40
5 9.42
6 6.28
7 3.14
8 15.70
9 21.98
10 34.54
11 50.24
12 125.6
13 100.48
14 628
15 150.72

Chapter 5.5
1 113.04 sq
2 28.26 sq
3 12.56 sq
4 3.14 sq
5 78.5 sq
6 314 sq
7 153.86 sq
8 379.94 sq
9 78.5 sq
10 1256 sq

TEST Chapter 5

1	113.4 sq"		18	F
2	28.26 sq"		19	T
3	78.5 sq"		20	F
4	12.56"		21	T
5	31.4"		22	T
6	100.48"		23	T
7	26"		24	T
8	22"		25	F
9	40		26	F
10	24		27	F
11	7 sq"		28	T
12	22 sq"		29	F
13	18 sq"		30	T
14	6 sq"		31	T
15	100 sq"		32	F
16	T		33	F
17	F			

Chapter 6.1

	Mean	Mode	Median
1	7.86	8	8
2	78.60	100	80
3	17.14	10	15
4	33.30	32	32
5	80.00	70	75
6	77.00	70	75
7	43.33	45	43.5
8	45.40	19	30
9	78.43	90	90
10	27.50	500	250

Chapter 6.2

1    $a=5$
2    $y=2$
3    $c=1$
4    $b=2$
5    $c=5$
6    $x=3$
7    $x=4$
8    $x=2$
9    $q=5$
10    $y=2$
11    $c=10$
12    $c=3$
13    $b=1$
14    $w=7$
15    $b=3$

Chapter 6.3

1    $x=2$
2    $a=1$
3    $y=3$
4    $x=3$
5    $q=5$
6    $r=3$
7    $x=5$
8    $q=4$
9    $y=2$
10    $y=10$

11    $y=2$
12    $r=1$
13    $x=5$
14    $x=11$
15    $y=2$
16    $x=10$
17    $c=11$
18    $b=7$
19    $r=7$
20    $x=3$

Chapter 6.4

1   40
2   .06
3   70
4   .7
5   7000
6   .011
7   .0017
8   220

9   .05
10   70
11   .0014
12   9000
13   300
14   .04
15   600

Chapter 6.5

1	508	9	122.000
2	0.390	10	453.600
3	91.000	11	0.700
4	8.045	12	3.784
5	109.400	13	131.280
6	3.105	14	5.902
7	4.719	15	0.910
8	26.400		

Chapter 6.6

1	10001_2	6	5
2	100000_2	7	3
3	1000001_2	8	15
4	1001_2	9	8
5	101_2	10	9

Chapter 6.7

1	1_5	7	41_5
2	13	8	24_7
3	50	9	45
4	32_7	10	10
5	1040_5	11	5
6	23	12	100100_2

TEST Chapter 6

1	y=3	11	55_5
2	x=5	12	13
3	x=5	13	95
4	r=7	14	66.43
5	x=3	15	75
6	y=2	16	90
7	b=2	17	30
8	q=5	18	11000
9	10111_2	19	0.06
10	10101011_2	20	0.4

ABOUT THE AUTHOR

Louis Parchman taught junior and high school mathematics in Duval County Florida (the 12th largest school system in the US) for 7 years including 5 summer school sessions. He also taught in several junior colleges in Jacksonville FL and Arkansas. He has 3 adult children and 4 grandchildren. He is married and lives in northeast Arkansas having retired after 40 successful years in the wireless industry.

www.ingramcontent.com/pod-product-compliance
Lightning Source LLC
Chambersburg PA
CBHW081237180526
45171CB00005B/449